# OIL ON THE BRAIN

To The Afflicted

# OIL ON THE BRAIN

The Yankees boast that they make Clocks,
Which "Just beat all creation."
They never made one could keep time,
With our great Speculation.
Our stocks, like clocks, go with a spring
Wind up _ run down again.
But all our strikes are sure to cause
Oil on the Brain.
Stock's per_ Stock's up,
Then on the wane,
Every body's troubled with
Oil on the Brain.

## Song AND Chorus

BY

# EASTBURN

Philadelphia J. MARSH 1102 Chestnut St.
New York S.T. Gordon          Boston O. Ditson & Co.

Ent. according to Act of Congress A.D. 1864 by J. Marsh in the Clerks Office of the Dist. Court for En. Dit. of Pa.

# OIL ON THE BRAIN

## Adventures from the
## Pump to the Pipeline

## LISA MARGONELLI

NAN A. TALESE / DOUBLEDAY
New York   London   Toronto   Sydney   Auckland

PUBLISHED BY NAN A. TALESE
AN IMPRINT OF DOUBLEDAY

Published in the United States by Nan A. Talese, an imprint of
The Doubleday Broadway Publishing Group, a division of
Random House, Inc., New York.
www.nanatalese.com

DOUBLEDAY is a registered trademark of Random House, Inc.

Parts of Chapter 11 previously appeared in *Wired* as
"China's Next Cultural Revolution," April 2005.

Frontispiece: Image of the song "Oil on the Brain" (on page ii)
is courtesy of Pennsylvania Historical and Museum Commission,
Drake Well Museum Collection, Titusville, Pennsylvania.

LIBRARY OF CONGRESS CATALOGING-IN-PUBLICATION DATA
Margonelli, Lisa.
Oil on the brain : adventures from the pump to the pipeine / Lisa Margonelli
p. cm.
Includes bibliographical references and index.
(alk. paper) 1. Petroleum industry and trade. I. Title.
HD9560.5.M3185 2007
338.2'7282—dc22 2006020789

ISBN: 978-0-385-51145-2

PRINTED IN THE UNITED STATES OF AMERICA

1 2 3 4 5 6 7 8 9 10

FIRST EDITION

*To my grandmothers, Alice and Helene,*

*and the wonderful big cars they used to drive*

# CONTENTS

# OIL ON THE BRAIN

# INTRODUCTION

My obsession with oil began at 2:15 on the afternoon of October 28, 2002. I was in Alaska's Prudhoe Bay oil field, standing on the edge of a tank holding 16,000 gallons of icy seawater, with a perfect pancake of spilled crude oil floating in front of me. I was observing an experiment on cleaning up oil spills in water, sponsored by the U.S. Minerals Management Service. The low arctic sun turned the snowfields lavender, and the oil equipment on the horizon was throwing pointy purple shadows. The chemist running the tests had made six spills already that day. He tossed a sandwich baggie of napalm on top of number seven and set it on fire.

At first only the napalm burned—like an industrial-strength tealight—but when the crude caught fire the flames began to dance. They blazed up violently, cracking and whirring as the hydrocarbon bonds broke. First a foot, then three feet, then higher. The flames were painfully hot, yet too brutal and fascinating to ignore—ten poli-sci lectures on the geopolitics of petroleum were soldered to my retinas on the spot. Oil the abstraction died and was reborn as a mythic molecule—powerful, violent, and charismatic—capable of running the world. With symphonic timing, the fire *whoosh*ed into a tall column above our heads as the remaining oil broke into heat, soot, and greenhouse gases. Then the flames drowned gracefully, and the arctic violet closed back in. I shivered.

This was my first direct encounter with crude oil. As long as I could remember, I'd been using my daily allotment of three gallons of gasoline (like the average American) but without giving it much thought. It was there when I needed it and more or less invisible the rest of the time. My understanding of oil was based in generalities. I had a vague working model of what oil was, where it came from, and how it got from place to place, but watching the burn gave me the sense that I'd been missing a lot by not knowing the specifics. *What is oil?*

For the chemist, the whole story was in the details. He poked at the dark waxy debris and said that for him oil was like fine wine, every field and every well holding an ever-changing stew of complex compounds, endlessly unpredictable and absorbing. He began musing about the components of crude, from the light gassy hydrocarbons to the heavy gooey ones: All of them have distinct personalities. The chemist imagined that the heaviest molecules would speak with the voice of Darth Vader.

He had oil on the brain. Now I, too, had oil on the brain.

Obsessions with oil are notoriously hard to shake. The title of this book, *Oil on the Brain,* comes from a song about the "oil fever" that gripped the United States shortly after oil was drilled out of a well in Pennsylvania in 1859. People rushed to boom towns with names like Pithole and Oil City, hoping to find their fortunes in crude.

My interest was different: I wanted to hear stories from the people who oversee oil's long journey to our cars. Americans talk about oil as a purely economic substance—we fulminate about its price, and most news about oil is printed in the business pages. I hoped to understand the culture of oil—the economies, destinies, and dreams shared by the people who live along the world's pipelines. In their stories, I hoped to make sense of the larger universe of oil.

My first glimpse of the culture of oil outside the United States came in the spring of 2001, when I reported on Saddam's sixty-fourth birthday party from Iraq. In Saddam's Baghdad, oil seeped into every aspect of life, trapping people in the money and allegiances it inspired. Oil smugglers were hanging around my hotel. Later that spring, I was sent to report another story in Arctic Village, a remote Native American community in Alaska's far north, where a dedicated group of a few dozen people has been fighting oil drilling in the Arctic National Wildlife Refuge for more than a decade. Flying northward to the village, I looked down at the silver Trans-Alaska pipeline threading its way around mountains and over rivers and was struck by how much it resembled a giant straw, sucking oil from the furthest frontiers of the continent to the suburbs and gas stations of the Lower 48. How did my life connect to the other end of the pipe? On that trip, I vowed to visit the wells at the head of the pipeline, and that impulse led me to the oil spill tests in Prudhoe Bay.

That cold October afternoon in Alaska, I didn't know where this would lead. All I knew was that I had the first symptoms of an oil obsession: jealous frustration. No sooner had I been mesmerized by the sight of a gallon of crude burning than another scientist said, "We did a spill once with 300 barrels of crude on ice. Flames 150 feet in the air. Burned for three hours. [Pause] You could see the fascination for an arsonist—those big flames." And soon after that I talked to a man who lit an even bigger fire with some of the oil spilled from the *Exxon Valdez,* and boy oh boy you hadn't lived, etc. However close I got to oil, there would always be someone who had been closer; whatever I learned, there would always be more. Oil was always right in front of me and maddeningly out of reach, a notoriously slippery subject.

To research this book, I decided to start with the one place I thought I understood: the gas station. In the summer of 2003, I began hanging out at a station in San Francisco, and I quickly discovered that it was a Rosetta stone for understanding my own conflicting emotional and economic rationalizations about oil. Nothing was as I expected. The one thing I thought I had a handle on—the price of gasoline, which is updated frequently and displayed prominently on large signs—turned out to be a chimera, albeit a fascinating one that reveals much about the behavior of American gasoline consumers and our role in the world.

From the station I continued up the pipe, farther and farther into the unknown. I rode in delivery trucks and spent time in a refinery. Hoping to learn how oil is created, and how it's drilled out of the ground, I headed to Texas to spend a week on a drilling rig. I became interested in the concept of energy security, and talked my way into the Strategic Petroleum Reserve. While doing that chapter, I realized I needed to visit the NYMEX oil market in New York. Along the way, people graciously helped me see their world, and I started to understand my own. Oil's price was only the beginning of the things I didn't understand. One of the most compelling characters in the whole oil universe turns out to be the American oil consumer, who ignores ever-higher prices and continues to use more oil month by month. The brute force of 194 million American pedals to the metal is changing the world day by day.

Interestingly, the American consumer was not an active part of the myth I believed about oil when I started reporting this book. Americans of all political persuasions tend to believe that we are trapped in an oil dependency over which we have no control. We are therefore victims of (fill in the blank): the Saudis, the Strategic Petroleum Reserve, OPEC, Exxon, GM, environmentalists who refuse to allow the building of new refineries, collusion between Dick Cheney and the oil industry, road-hog

SUV drivers, China's growing economy, and so on. Even the president, himself a son of the oil industry, has characterized American oil dependence as an "addiction," as if it were a tragic product of brain chemistry rather than one made of daily economic choices. Not only are these conspiracy theories useless for examining our relationship to an economy as complex as that of oil, they actively discourage strategic planning by suggesting that all circumstances are beyond our control. This uncurious anger and passivity among consumers, I've come to believe, is the real oil conspiracy.

Overseas, where the United States now buys the majority of the oil we use, I tried to find the forces that will really influence the flow of oil in the future. I visited first America's oldest petroleum partner—Venezuela—which is involved in a massive experiment to reinvent the oil state. Then I went to the newest oil state—Chad—where Exxon and the World Bank have started a different, but still massive, effort to reinvent the oil state. Obviously, oil states are troubled, but whether they can be reinvented, and whether the relationships between these nations and America will survive the process, is uncertain. In the Persian Gulf, I went to Iran, formerly a U.S. ally and now an enemy, to revisit the scene of a strange one-day military battle that changed the role of the United States in the Middle East forever. In Nigeria I visited with a warlord who changed the world price of oil with nothing more than a cell-phone call. In my travels I worked away at the knot between American oil dependence and foreign policy, military involvement, and our peculiar brand of empire. And finally, I went to China, where the world's newest oil consumers are designing the cars and fuels of the future.

**The years between** 2003 and 2006 were particularly tricky years to try to make a portrait of the oil economy. During that time, the price of a barrel of crude nearly tripled—and that was only a code for thousands of other interconnected changes. Some of the changes were deliberate: When the United States invaded Iraq, it set off a series of repercussions everywhere from Iran to Russia. Some changes were historic: China's and India's economies took off, requiring huge amounts of fuel. As I write this, the entire world of oil is reorganizing in a way I did not imagine when I first watched crude burn in Prudhoe Bay.

The changes had already started on that October day in 2002, and the sudden arrival of the polar bears should have been a clue. Everywhere I went inside the Prudhoe Bay facilities, there were posters featuring a photo of a polar bear's head and the words: "Danger! An unusual number of polar bears have been sighted near the facilities."

The story was that an ice floe with ninety-nine polar bears pulled up in the Beaufort Sea. The polar bears got off the floe to hunt around. But then the ice left without them, stranding the bears like tourists who'd missed their cruise ship. At every break, the team leader reminded us to look carefully at the snow to see if something thirteen feet tall and white on white was scurrying in our direction.

That was the warmest October in memory. One day the temperature approached the melting point, 25 degrees higher than it was supposed to be. Even the skeptics began talking about global warming, and by 2006, as I write this, few would deny its effect. The polar bears are part of a system newly in flux, dislodged by the carbon in the very hydrocarbons that lie beneath the North Slope, wandering around in a changing landscape. Beneath the polar bears, the subterranean landscape of Prudhoe Bay was also changing. In August 2006, BP announced that it was shutting down the oil field for repairs. While the public's attention was turned elsewhere, America's largest oil field had faded into near-irrelevancy. Now it produced less than 3 percent of the country's daily needs, and its output was so small the company hadn't cleaned some pipes in more than a decade.

In the midst of all of this change, I have felt a bit like one of those disoriented polar bears, wandering around trying to get the lay of the land as it shifted under my feet, but that's a feeling we'll all have to get used to. Every second, the United States burns another ten thousand* gallons of gasoline: We are driving ourselves into a new world.

---

* There are forty-two gallons in a barrel of oil. Oil is more often counted in barrels than in gallons—a throwback to the 1860s, when it was sold in handmade wooden casks. Back before pipelines, oil drillers near Pennsylvania's Oil Creek loaded their oil into barrels and stacked them on barges. They'd dam up the creek and try to float the barges downstream on bursts of water. Often enough, the barges crashed, and photos show children scooping oil from the creek in buckets. Things have changed since then. In July 2006 the United States consumed approximately 20,656,000 barrels a day. Imports made up 59.8 percent of the total.

# 1 GAS STATION  *CHASING THE HIDDEN PENNY*

## Regular Unleaded $1.61⁹⁄₁₀

Twin Peaks Petroleum sits at a welcoming angle to a busy San Francisco intersection. On this morning in the summer of 2003, a thick fog has crawled over the station, folding each of the eight drivers standing at the pumps in an envelope of cold mist. At the back of the lot sits a garage where a small convenience store glows. On the storefront is a poster of an ebullient snowman clutching a cola, while icicle letters drip the words COLD POP over his head. Inside the convenience store, among the security cameras and parabolic mirrors, the Doritos, cigarettes, and Snapples, jammed into a space no larger than a postal truck, a tall man with dark circles under his eyes appears to doze. His eyelids hang low, twitching; he mumbles; he moves with excruciating deliberation as he counts change.

I am leaning against a shelf holding several grades of motor oil, individually wrapped strawberry cheesecake muffins, and four flavors of corn nuts: picante, regular, nacho cheese, and ranch. I am no more lively than B. J., the droopy manager. And I am recording the flavors of corn nuts in my notebook to stay awake. "Corn Gone Wrong" say the packages. I record that too. I've come to the gas station to watch Americans buy gasoline, as a way of understanding how we fit into the trillion-dollar world oil economy. But now that I'm here, I realize I've been here before, bought gas so many times myself I feel there's nothing to see. The fumble, the stuporous swipe of the card, the far-off look: I know them well. Gas stations are everywhere, but when you're in one, you're nowhere in particular. Icicle letters are taking shape in my head: WHAT DID YOU EXPECT?

I keep writing: Trojan spermicidally lubricated snugger fit, two Sominex and a folded paper cup, phone cards, batteries, air fresheners printed to look like ice cream sundaes, a greeting card with a picture of a pansy and the words "You're too nice to be sick." The customers standing out at the pumps have a preoccupied, anxious look—could they be dis-tracted enough to buy a card that says "You're too nice to be sick"?

**Gas stations** are collections of incidental items, impulses, and routines that seem in themselves to be inconsequential but aggregate into a go-liath economy when multiplied by the hungers of 194 million licensed American drivers. Corn nuts, for example, are part of $4.4 billion in salty snacks sold at gas station convenience stores yearly, nearly all impulse buys. The hopeful purchase $25 billion in lottery tickets. People with the sniffles spent $323 million on cold medicine at gas stations in 2001. And the faint smell of gasoline near the pumps? In California alone, the amount of gasoline vapor wafting out of stations, as we fill our cars, totals 15,811 gallons a day—roughly the equivalent of two full tanker trucks.[1] In the gas station, we've collaborated to create a culture of speed, con-venience, low prices, and 64-ounce cup holders, which allow us to ex-press what the industry calls our "passion for fountain drinks." Japanese auto executives have hired American anthropologists to explain the mystery of why the purchase of a $40,000 car hangs on the super-sizing of the cup holder.

And then there is the gasoline: 1,143 gallons per household per year, purchased in two-and-a-half-minute dashes. We make 16 billion stops at gas stations yearly, taking final delivery on 140 billion gallons of gasoline that has traveled around the world in tanker ships, pipelines, and shiny silver trucks. And then we peel out, get on with our real lives, get back on the highway, or go find a restroom that's open, for Pete's sake.

With a wave of our powerful credit cards, American drivers buy one-ninth of the world's crude oil production per day. That makes us ele-phants in the global oil economy—our needs are felt around the world, from the tiniest villages in Africa, the Amazon, and the Arctic, to the highest towers in Vienna, Riyadh, and New York. When we lick our lips, they open their taps. When we are in a funk, their governments fall. Here in front of the pump, surrounded by buntings in the joyful colors of chil-

---

[1] Four out of five people feel, on a gut level, that gasoline is toxic, according to one study. A majority dislikes the smell. (A contrary 11 percent say they find the scent "pleasant." More on them in Chapter 4.)

dren's birthday party balloons, we have the opportunity to be our truest selves in the great, over-the-top drama/business that is the world oil supply chain.

But as you know, buying gas can be done by the living dead. Swipe card, insert nozzle, punch the button with the greasy sheen: Gasoline flows into the tank while money flows out of the bank account. Filling a car seems less like making a purchase than a ritual, a formality that isn't quite real.

It's not even clear what we're buying—gasoline's fantastic uniformity means one is as good as another. Water doesn't mix with gas, so beyond occasional traces of vapor, we don't even have to worry about buying substandard gasoline. And all traces of where the fuel came from are completely erased by the time it gets to a gas pump. Texaco gasoline is no longer from Texas, and gas from Unocal is not from "Cal." Both companies have been purchased by Chevron, anyway. If gasoline were coffee, we might believe the Baku blend offered a fast but mellow ride.

As if acknowledging the futility of trying to stand out from the pack when 168,987 gas stations are selling essentially an identical chemical mix, stations have adopted a clannish ugliness. Whether they're in Fairbanks, Alaska, or Pine Island, Florida, they all subscribe to the familiar topography of canopied islands, cheerful plate glass, struggling hedges, and "Smile. You're being watched by a surveillance camera" signs. Predictable they are, to the very last 9/10ths of a cent, which is permanently printed on every last gas price sign in the land.

The gas station's blandness is misleading, though. Hidden in its windows, pumps, and hedges are clues to the true nature of the American bargain with gasoline and the enigma of its role in the world.

**On the counter** in front of B. J. stands a line of purple plastic wizards, stomachs filled with green candy pebbles. Their shiny eyes stare at me expectantly.

At the periphery of my vision, a van enters Twin Peaks yard and parks near the fence. In the time it takes the door to slam, B. J. grows a foot taller, loses his paunch, and becomes a man of action. He snaps the countertop open, bounces into the yard, and lands in front of the van driver in one tigerlike swoop. Words are exchanged. The driver sulkily returns to his van and B. J. returns to the store, shaking his head. People try to ditch their cars in the station and take the bus, he explains, taking his position behind the wizards. "The customer is always right," says B. J., "but bad people going round."

Like vapors, bad people always seem to be wafting through the gas station. Last week B. J. ran out to stop a truck that was barreling toward the station's lighted canopy. The truck driver ignored him and crunched the canopy. Cars have driven willy nilly through the hedges as he watched. Nightly, people break through the chains on the four entrances. Once he found a gun in the hedge, stashed by a kid on the way to juvenile court. His response? Shave the hedges. Every morning he cleans up garbage, cans, and bottles filled with things we won't discuss. Daily, and constantly, people try to steal: window squeegees, sodas, condoms, money, and phone calls. Behind B. J.'s head are the counterfeit $20 bills the station has intercepted.

People use elaborate schemes to steal gas, he explains. Sometimes they'll pay for $5 and shut the pump off when it reaches $4.75. Then they return to the clerk, telling him to turn the pump back on, knowing that the pumps don't turn off after dispensing amounts less than a dollar. Then they fill their tank and drive off. Others play on the sympathies of the attendant or accuse him of trying to cheat them. In stations where people pump before they pay, they often just drive off. The average gas station loses more than $2,141 a year to gasoline theft. Some lose much more.

Think of a gas station as a crime scene before the fact, and you'll start to appreciate it as a maze engineered for belligerent rats. Hedges, which I'd interpreted as a pathetic attempt at dignity and baronial pretensions, actually eliminate escape routes for would-be robbers, limiting holdups. Many convenience stores buy "target hardening" kits, which include decals imprinted with rulers so that clerks can tell the police how tall the robbers were, two stickers that say "No 20s, no 50s," two "Thank You" decals, and one "Smile. You are being watched by our video security."

Even so, crime is always evolving. "After we did target hardening in stores in the 1980s, the crime moved to the pumps—carjackings and abductions," says Dr. Rosemary Erickson, a sociologist who's studied gas station crime for thirty years. "Now it's public nuisance crimes in the parking lots. Gas stations are considered a magnet."[2] Nearly nine percent of U.S. robberies happen in gas stations and convenience stores, and the average gas station lost $1,749 to robbery in 2004.

Some of the crimes are not about money at all; they're about free-floating anger. When gas prices are high, more people get "pump rage" and try to drive off without paying for gas. The Indian and Pakistani immigrants who own and staff many stations bear the brunt. After 9/11, people who were angry at some vague combination of OPEC and Osama bin Laden attacked a hundred clerks at 7-Eleven gas stations and convenience stores in a month. Five men were killed for looking "Middle Eastern." A photo of bin Laden in a 7-Eleven uniform circulated on the Internet. The National Association of Convenience Stores issued a list of tips to discourage customers from attacking employees—including posting flags near the cash register.

In this harsh microclimate, B. J. maintains an impartial vigilance. When a woman enters and asks to use the phone, he activates the indifferent stupor, gesturing vaguely in the direction of a pay phone.

---

[2] Just about the only bad things that haven't happened in gas stations are the things we think happened there. Urban legends about AIDS-tainted hypodermic needles supposedly hidden in nozzle handles have flown around the Internet for years. There's never been a legitimate newspaper account of that happening, though it fits with a larger sense that "anything" could happen in a gas station. Stories about cell phones allegedly igniting fuel at the pump are also apparently without basis in fact.

More alarming, though, is what *has* happened. Between 1997 and 2004, at least 163 people were hit by flash fire while fueling their cars. One person and one dog were killed when the fuel was ignited by static electricity (often caused by sliding across car seats in winter clothing). Reports of these stories and others can be found on the Web site of the Petroleum Equipment Institute, www.pei.org/static/fire_reports.htm. In January of 2002, a customer driving a Ford 150 experienced the following:

*Customer paid clerk $20. Started nozzle pumping and got back into truck because it was cold. She noticed pump had passed $20 and jumped out to grab nozzle to stop gas. Vapors burst in a big fireball.*

In May of 1996 a customer driving a Honda Accord wrote:

*My son went from the rider's side of the car, opened the fuel door, and unscrewed the cap. Took the nozzle from the dispenser and put it in car. Before dispensing began, flames shot out like a dragon shooting out fire.*

Though most pumps have been modified to avoid the fires, it's probably worth grounding yourself by touching your car's door handle before touching the gas nozzle.

In a less charitable person, working in a gas station would have long ago brought on moral exhaustion and cynicism, but B. J. has created a worldview that embraces the station. "Everything is in the gas station," he says. "Good people and bad people are here. Not many honest people."

Spending mornings with B. J. gives me time to reconsider the extreme disinterest I've noticed in convenience store clerks over the years. B. J. is clearly operating the selective stupor on a very high level— perhaps as a watchful hibernation, or a trancelike sensitivity to the station's periphery. Anyway, he has perfected the art of being aware of the bad people going round without letting on that he's awake.

A man comes in carrying a tiny dog and asking for M&M's; B. J. gives him a big grin.

A woman asks for the restroom. B. J. waves noncommitally at the sign, "No Restroom," and the woman pixillates back into the fog.

When a gnarled old man with a froth of white hair and huge yellow shooting glasses hands him a $10 bill, B. J. says gently, "Hello, my friend, how are you?"

However harsh the gas station microclimate is, it's also a neighborhood for anyone who wants one. One industry focus group was surprised to find that customers had "deep feelings" for clerks. "Convenience store clerks are at the bottom of the retail ladder," says Jay Gordon of industry publication *C-Store Decisions*, "but that's not how people perceive them. One woman described her local clerk as a "superhero, with six arms, always giving directions, napkins, and keeping things moving. Superhuman."

One morning B. J. talks about how he left his family's farm in India in 1984. His family feared he'd be persecuted because he was Sikh, and the family's oldest son, so he left the country. He went first to Singapore, then Malaysia, Australia, New Zealand, Canada, and finally New Jersey, where he worked in a relative's gas station. His English is haphazard. "New Jersey. My skin no like cold. Look ugly. My ears turn black." He grins at his ghoulishness. "I pack up me money and I come here." California. His family has joined him here. His daughter is a nurse, his son an X-ray technician, and his son-in-law an engineer who's getting his MBA. Back in India, his younger brother runs the farm. B. J. has a theory about gas stations and life in general: "If I am good all is good. If I am bad all people look bad. I am nice and I have nice kids. If I am bad they will

steal. . . ." He explains this to me a few times until it takes on the attri-
butes of a philosophy.

The customers keep coming. They buy energy drinks and power bars
and candy and Visine: Everyone spends time in front of the coolers,
which are called "the vault." The name properly recognizes both their
position in the store—always opposite the door—and their role in bringing in a high percentage of the store's profits.

Snapple is advertising a drink involving bananas. The decal on the
cooler says "Release Your Inner Chimp." I look at the stream of distracted
people coming into the store and wonder how many of them are attracted by the faux naughtiness implied by "Corn Gone Wrong" and
"Release Your Inner Chimp." Why is it that the coolly rational customers
who shop ruthlessly for the cheapest gas turn into formless emotional
mush—susceptible to the likes of their "inner chimp"—when they enter
the convenience store?


**"I make more** money selling water than gas," says owner Michael
Gharib. "And the gas gets shipped around the world and goes through a
refinery and still my customers want it cheap." Wearing a pressed pink-
striped Ralph Lauren shirt, Gharib arrives at the station ready to spread
his sense of order, compulsively arranging the mints as he talks about the
twenty years he's owned the place. When he started he made his money
selling gas with a profit margin of 10 cents a gallon, but as gas margins
have fallen, he uses the store's 25 percent margin to boost the overall
business. He moves on to dust the Daffy Duck Pez dispensers, straightens
the STP carb cleaner, and leaves the dusty Fritos in a cup alone on the
bottom shelf. Onward to the Skittles! Gharib has the shoulders of a
weight lifter, and big, liquid brown eyes. When he's in the store it seems
suddenly smaller, its two hundred products ajumble until he personally
straightens them. As I watch him, it dawns on me that the thing I've
thought of as a gas station for the past twenty years is actually more of a
Skittles emporium that sells gasoline.

The candy wizards, for example, contain about 50 cents profit. To
make that 50 cents selling gasoline, the store would have to sell 10 gallons of gas. (And presumably the wizards are not combustible.) Nationally, sunglasses have a 100 percent markup, ice is 60 percent, candy is

43, and cigarettes are 19. Gasoline's profitability has been falling over the past few years, reaching just 7 percent in 2004, the lowest in twenty years. It seems that the more ruthlessly we shop for the cheapest gasoline, the more vulnerable we are to the likes of Corn Gone Wrong: Impulse buys make up three-quarters of the $132 billion spent in convenience stores. The gigantic economy of oil marketing comes to one ironic point: Selling gasoline in America requires the assistance of candy wizards.

**Running a station** is a tough business: One in six gas stations has closed in the last ten years. Gharib has a competitive edge, an attention to the details and balance sheets that has stood him well.

In the early 1980s Mobil owned the station. Gharib, a mechanic, leased it from the company. He had to sell Mobil brand gas but he got a guaranteed profit of 10 cents a gallon, and he ran the garage as a business. He never had to worry about gas prices, maintaining the pumps, or whether the tanks under the station met the latest code. A phone call to Mobil took care of virtually everything. In return, Mobil, like other brands, depended on men like Gharib, who had relationships with their customers, and their cars, to draw in business and sell more gas. Gharib, who was born in Iran and is fluent in American culture, was probably a good draw—he's friendly, warm, and shrewd. Let me put it this way: His wife is one of his former customers. But the business of selling gas changed. Mobil stopped spending money to keep up the site and then sold it to another brand, which also didn't put much money in. In the early 1990s the station had just two old pumps. "The brand didn't see the money in this location," says Gharib, "but being here ten years I knew the potential—it's convenient, and half the customers have been coming here for years and years. Firemen, police. Middle class."

Gharib swung a deal to buy the station, remodeled it, and named it Twin Peaks to make it part of the neighborhood. (He and his wife designed the logo on their honeymoon in Fiji. "There was nothing to do but sit on the beach and scuba dive," he says. And think about gas stations.) He reopened it as an independent, selling unbranded gasoline at discount rates. As the market got tougher, he added the convenience store. Eventually he leased the garage to another mechanic.

Gharib has agreed to let me hang out in his station because he sees himself as part of a brotherhood of independent station owners. Independents own about 35 percent of the stations in the United States, but they are overshadowed by the major oil companies with their big flashy brands and advertising. When I called Gharib and introduced myself, he was amused by my request, but agreeable. "There's not much to see," he cautioned, adding that he won't talk with TV reporters anymore. "They're really interested in us when prices are screamin'," he says, but when the interview appears on the news, they've lost all the nuances. "It's totally different," he says. "It reflects badly—not on me personally, but on the whole industry."

Independent stations see themselves as underdogs. They buy gasoline from wholesalers called jobbers, who buy wholesale gas on the spot market and truck it to them. The gasoline they sell is chemically identical to that of the branded stations—it comes from the same refineries, travels through the same pipelines, and sits around in the same tanks. But while the brands advertise that their gas contains special detergent formulas, the independents all use a generic formula and discount the wholesale gas by 2 to 3 cents per gallon. Without national advertising or a strong image, independents often try to keep their prices lower than the branded stations, which means they have to skillfully navigate both the wholesale market and the retail market. They're wily and willing to take risks, and they need a reasonably large line of credit. "You can make a lot of money," says Gharib, "but sometimes you make negative. When the refineries sneeze we're vulnerable."

Gharib invites me back into to his office, a tiny cinder-block room containing a metal desk and two metal chairs. He sorts his mail between the minarets of a plastic Taj Mahal on the desk. The space under the desk is filled by three and a half sacks of quick-setting concrete mix. Stacked on a shelf by his left ear is a pile of video decks and a TV for the surveillance cameras. There is a folding chair on the right side of the desk, but it has a gas nozzle on it.

He points to the end of the nozzle that connects with the hose. "Break-away valve," he says. "Separates at twenty pounds of pressure." If somebody drives off with the hose in the tank, Gharib can just replace the nozzle and break-away valve rather than having to replace the whole pump. Twice a month somebody drives off with the nozzle in his or her tank. "Friday. Hot. Nice weather. You're talking on the cell phone and the kids are driving you nuts. . . ." He shrugs. Some drivers are embarrassed and try to pay immediately, but the majority insist it wasn't their fault and some claim they don't have insurance. In the gas station, something seems to tempt good people to act bad.[3]

Gharib pulls up a eadsheet on his computer to show me how the price of gas is constructed. Today $1.61 and 9/10 is the price for regular. Of that, 37 cents a gallon are the state and federal fuel tax and superfund taxes. Sales tax costs another 13.3 cents, and when the price rises, the tax goes with it. Visa and MasterCard make 3.8 percent on every card purchase, which make up seven out of ten fill-ups. Then there's the property. "Something as basic as this"—he waves at the cement walls—"is $1.5 million for the property and another half million for the station. You carry a mortgage on that and pay additional tank fees and maintenance." And then there's wholesale price and overhead: That puts Gharib's margin at 5 to 6 cents a gallon today, which isn't a lot, but with his low prices he does high volume.

Gharib believes he needs to keep his prices approximately 10 cents lower than nearby brands to attract customers. When prices are stable, that's easy—there's plenty of gasoline on the wholesale market, and he can easily stay under the brands and still be profitable. When prices are climbing, though, Gharib is in danger—he needs to lower his profit margin quickly to stay competitive with the brands, which are cushioned against price rises by the refiners. He needs to be able to make a quick decision. Should he fill his tanks at today's price, assuming tomorrow's will be higher? Or should he wait until after midnight, hoping tomorrow will be lower?

When prices are falling, he's in worse shape because he needs to be sure that he hasn't filled his 8,000-gallon tank with expensive gas that he has to sell at a loss to be competitive with the brands. What's worse, when prices climb steeply, it's usually because the refineries don't have much fuel, and then the wholesale gasoline price can be 10 cents more

---

[3] Long ago, gas station designers gave up on the idea of changing our behavior and decided to focus on minimizing damage. Throw your car in reverse and back over a pump: You will not end up in the tower of flames you see in the movies. Instead, the top of the pump will flop over and hidden valves will shut off the gas at ground level.

than the branded gas. That's an inversion. "If you started your business in an inversion, it really scares the heck out of you," he says. "You can lose thousands of dollars in a day." Gharib never takes days off, he says, and if he's with his family, his cell phone is always on. "I imagine my retirement floating out there like an oasis"—he smiles—"and I'm getting into real estate."

**Small businesses** aren't for scaredy cats, but being an independent gas dealer is particularly rough because you're working in a market dominated by some of the biggest corporations in the world. When the major brands sell gas through their dealers, they maximize profits by selling to some dealers at high prices and to others at lower prices. Sometimes a station near a freeway entrance will be 15 cents higher than the same brand on a busy corner with competing stations nearby. This is called zone pricing, and refiners determine prices for different stations based on the neighborhood, competition, and traffic volume, among other things. (Though they don't like to talk about zone pricing or their criteria, it seems clear that they create a computer analysis of every station in their chain and assess the competition, to create pricing models.) This system is controversial—and it makes some dealers and consumer groups furious. An investigation by the Federal Trade Commission in 2000 described zone pricing as "an earmark of oligopolistic market behavior," but by 2004 the commission concluded that it was okay, because while it hurts some consumers, it helps others. If prices are too high at one station, you can always go elsewhere, it reasons.

The retail gasoline market is tremendously complicated. One thing that analysts and the Department of Energy agree on is that independent gas stations help keep gas prices lower by competing with the brands. However, zone pricing gives brands a competitive advantage because they can choose to reduce prices at some stations so that nearby independents cannot compete. Where brands drive out competitors, consumers probably will have to pay higher prices. Because the market is so complex, and constantly evolving, there's a fear that regulations to limit the power of brands may end up backfiring and hurting consumers. For example, forcing stations to sell gas with a minimum markup, which is the law in eleven states, may end up punishing consumers because it prevents big discount chains from opening "hypermarts," which sell high volumes of gas at very low prices.


**Gharib says** he's squeezed by both his customers and the refineries, but he can't resent either one. "Everybody's pissed at the oil companies," he says, "and the refineries are probably making a killing. But if they don't make money we'll be out another refinery. Then what will we do? Ride our bicycles?"

Gharib says that environmental regulations hit independents worse than the majors. California's requirement that stations replace their underground tanks in the late 1990s put many independent gas stations out of business. He ticks off some of the other certifications and tests required by the federal, state, county, and city governments. "I used to have just two certificates," he says. "Now I have filled up three frames with these things. With the last two I just taped them to the wall."

Above Gharib's desk is a bank of video decks and a small TV, which shows the station from different angles. "This is probably the best investment I ever made," Gharib says, craning his neck at the TV. "Twelve thousand dollars. I've got eight cameras and sound too. You have to have this."

In the industry, employee theft is called "merchandise shrink," and it apparently costs an average of $11,378 per store. That seems impossible here—who would want to steal that many Dr Peppers? One manager before B. J. stole $70,000 over the course of a year and a half, Gharib says. Another employee came back and held up the store. "B. J. is really the only person I trust," says Gharib.

Sitting in the little cinder-block office, I'm not sure why anyone would agree to sell gasoline; between rude customers, ruthless competition, and pilfering employees, you could be broke or dead in an instant. Gharib laughs. "I've always got an exit strategy." But then too, this business is so difficult it's fun, and maybe that's what makes it attractive. Whenever he thinks about leaving, he has another idea for making money.

Just last week he took a critical look around the store and decided the greeting cards had to go. He's replacing them with an ATM, which will yield a profit of about a dollar a transaction—similar to selling 20 gallons of gas, except that you don't have to worry about an ATM catching fire, leaking into groundwater, or evaporating.

**Once I started** hanging out in gas stations, I saw their skeletons everywhere. My drives around town became expeditions through an old gas station battlefield. One near where I live is now an Ethiopian restaurant. Steam from cooking fava beans curls out of the old office. Farther down the same street is the hulk of a Texaco station—the green and white tile gives it away—which now hosts a cactus nursery. Another is a mattress discounter. They are the last living reminders of gasoline wars long gone by, and they tell the story of how the industry has adapted.

The first gas station opened in 1907, and the Model T Ford came out the following year. From then on, oil companies were in hot competition for customers. Refineries were expensive to build and relatively cheap to operate, so oil companies focused on building powerful brands and occupying more land to sell more gas. They played for market share rather than profits. This same somewhat dysfunctional business strategy continued until 1973, when the Arab oil embargo changed the gas game for everyone.

In the early days, stations spread along roads and highways like weeds. In 1920 there were 15,000 stations in the United States; by 1930 there were 123,979. In one six-week period in the 1920s Shell threw up 100 identical gas stations between San Jose, California, and Santa Barbara—one station every 2.6 miles. Early stations were designed to look familiar and substantial, like cottages or university buildings, but they gradually developed ideas of their own. Frank Lloyd Wright saw stations as the "embryo" of a "well designed, convenient neighborhood distribution center" in one of the utopias he designed.[4] If you forget the part about utopia, his vision has been achieved in every local 7-Eleven. Gas stations were the seeds for roadside sprawl, incubators for our peculiar highway culture of a "home away from home" everywhere you go.

---

[4] Wright's plan called for the gas to come down in hoses from above, reminding historians of one of his favorite images: the cow. "Could it be that the famous architect saw petroleum—which he lyrically described as 'the wealth of states, the health of nations'— as the machine age counterpart to the cow's milk?" speculates Daniel Vieyra, in his book *Fill'er Up*. "As Wright asked, 'How many trusties and lusties besides her lawful calf have pulled away at her teats these thousands of years?' "

And as these gas stations snaked out across the land, they all shared the same problem: Gas is gas. Why should I buy yours? No matter what additives you put in, and what you claim they do for the engine or the driver's self-esteem, gas is still just gas. Stations had to deliver more: sleek white tiles, restrooms purportedly inspected by nurses, free silverware, attendants in green go-go boots. Stamps! In the late 1960s the oil industry dropped $150 million on trading stamps, a fantasy money that could be traded for prizes in an "idea book." Driving down the street then was a drive through a dreamscape—each gas station was a slightly different theme park, with slightly different stamps as currency, and very similar gasoline. Perhaps it was the illusion of choice that counted. To create loyal customers, the oil companies invented credit cards, which then took on a life of their own. Whatever the station, the driver was always the star. By 1972, entering most of the country's 226,000 gas stations got you lots of attention. Fill 'er up? Want your window washed? Your oil checked? Double stamps?

The fanfare of buying gas is gone now, and so is the fun. On the freeways I drive past flat, wide canopies holding up broad bands of color to signal the brand they sell and a giant sign to show the price. Modern gas stations are designed for big corporate mergers: The colors and brand IDs can literally be snapped off and replaced by a new owner overnight. And studies have shown that a surprising number of gas buyers can't even tell what brand they're buying when they're standing at the pump. Customers are left to fend for themselves. In exchange for pumping our own gas, checking our own oil, washing our own windows, and generally wallowing around at our true place in the food chain, we save a few cents. There is no man who "wears the star" to trust our cars to.

The man who created a business selling self-serve gasoline in Northern California and much of the West is Herb Richards. Herb didn't invent self-serve gas—credit for that usually goes to an entrepreneur named Frank Urich, who opened a station in L.A. in 1947, with girls on roller skates collecting money from customers. But Herb was part of the generation of men who figured out how to sell gasoline to the masses after World War II, creating a culture that remains, in some form, to this day.

I found Herb at his office, which used to be on the outskirts of San Jose but is now in the midst of miles of sprawl. "We had ten years before people caught up with us," he says about the self-serve chain he co-founded. "Fifteen years to Chicago and twenty-five to Boston. They still don't do it in New Jersey."

Herb is tall, spare, and forceful, in a gentleman-cowboy sort of way. He greets me wearing a pale blue tailored suit. He is ninety-three, and he still thinks it's fun to sell gas. Part of what makes it fun is that he doesn't sell gasoline to cars but to humans, who have simple emotional machinery. I ask him about the stamps. "Ninety percent of women loved stamps and 100 percent of men hated stamps," he says. Herb doesn't stoop to dramatize his stories, or brag. He doesn't need to. "We decided that women controlled the money. Rather than give a lower price on gas, we'd give ten times the stamps so the men with the stamps would be well received by their wives." Why should you buy our gas? So your wife will love you. Herb has pool player's eyes, assessing the angles, and the cool detachment to see that gas stations sell ideas, not gas.

Herb fell for oil at fifteen, when a chemist at the refinery where his father worked showed him the things he created from crude. "Cold creams and such. It stuck with me." In the Depression, he worked his way through college in a gas station, before getting a petroleum engineering degree at UC Berkeley. During the war he did petrochemical research at Stanford. To make money on the side, he built a little refinery to recycle motor oil and worked with his dad's gasoline distribution business. He even had an oil well in the mountains near Santa Cruz that produced 10 or 20 barrels a day. "It was kind of fun," he says. After the war there were many good years of selling gas. Three stations. Six. Fifty. A hundred and twenty. California, Nevada, Arizona, and Oregon.

Eventually Herb made a lot of money—enough that people speak about it in a whisper—but he retains a tough, Depression-forged skepticism. His desk is from the 1950s, with a wicker inset and brushed brass cylinders for front legs. His office is austere, with old wood paneling and a low ceiling. Behind Herb's sharp head hangs a photo of him shaking hands with Margaret Thatcher. "I've always admired her. She's a very gracious woman. She's tough. Very fair. I wish she was president."

After the war ended, Herb and a partner drove to L.A. to look at Urich's first self-serve gas station, which had rows of pumps. Herb doesn't mention the girls and the roller skates; what impressed him was the volume of gas the station sold. A self-serve station could sell four times as much as a full-service station. In 1947 he and his partner opened the first self-serve in San Jose, selling unbranded gas. People liked the idea of saving money, he says, but the station never actually posted the price, which was around 30 cents a gallon. What attracted people was not the price but the concept of thrift. "We *imaged* selling cheaper than other people. Big gaudy sign: SAVE. It wouldn't be effective now—people are very conscious of price." More stations.

Selling four times the gas required four times the customers, but without service and "special" additives the brands were offering, Herb and his partners needed to offer more. Eventually they settled on Cadillacs.

They gave away dozens of Cadillacs. Cadillacs with sharklike bumps on their hoods, chrome missile tips on the grille, and aggressive rear fins. Every other Thursday he raffled off a car—an eight-cylinder dream machine. The raffle was broadcast over the radio, and the ticket stubs were scrambled in a clean cement mixer. Customers in five states gathered together the tickets they'd gotten when they bought Herb's gas and sat by the radio to see if a Cadillac was in their future. "Cadillac gave us a deal." Herb shrugs. "It might have cost us a penny a gallon, maybe 2 cents."

Because the gas Herb sold had no brand, he was free to create them. "Regal. Mohawk. Beacon. Too many names." What he didn't invent, he borrowed freely. Driving in Nevada, he came upon a station with the unappealing name of Terrible Herbst and recognized that the unpleasant is memorable. "It's a name no one likes but no one forgets." With a partner named Robinson, he opened a new chain named Rotten Robbie. "Everybody knows where Rotten Robbie is whether they like it or not. It's interesting how that theory works."

In the early days every single station was a battle. "I thought it would grow big. I think I was optimistic. We had to fight our way through every city council because self-serve was against the law. We weren't just building stations; we had to argue why they were safe. There wasn't anyplace we didn't fight." Once the stations were built, towns fought their big gaudy signs.[5]

---

[5] The struggle for self-serve gas was epic in a spaghetti-westernish way. Hugh Lacy, the vice president of Urich's original station in L.A., received forty death threats and was run off the road by six men "who took off when I produced a .38 automatic." Stations were attacked, and Urich himself was shot at while driving a gasoline truck. Writing to a gasoline historian in 1971, Lacy said, "You have missed entirely the romance of this industry."

And always there was the gas war raging in the background. During the 1950s, the major brands battled against each other, nowhere more viciously than on the West Coast, where growth was fastest. While fighting each other, they all fought Herb. Refiners didn't want to sell to his stations. "The competition could see what it was doing to the marketplace so they fought it. God, that made life interesting. When someone tells you you can't do it." He shakes his head, but not ruefully. "It's the boy in me."

To get around the gasoline blockade, Herb started using a set of exchanges with independent refineries and other dealers. Exchanges are based on the idea that all gasoline is identical, so companies can exchange the gas "on paper" between cities, without having to truck the gas. So if Herb gave gas to, say, Eagle's stations in San Jose, Eagle would bring gas to his stations in Los Angeles, which annoyed the L.A. refineries that refused to sell to Herb.

When the major brands in the area tried to underprice him, he responded with stunts like giving away a free gallon of gas to any voter who showed up with a ballot stub on election day. "I tried to tell Shell we were being patriotic. You can't make 'em too mad."

The only way to fight big was to get big himself, so Herb's partnerships pushed into new businesses. Tired of struggling for tanker space, Herb and his partners started their own jobber/hauling company to supply the stations. When the majors wouldn't let him store gas in the far north of California, they bought an old coast guard station, complete with a barge dock and fuel storage. "Thirty days later the majors said they'd give us a contract. You had the struggle. And it was amazing how easy things became when you were in the good graces of a major."

The great postwar boom in autos, highways, and suburbs rolled out on a wave of cheap gasoline, and men like Herb shaped a culture that seemed uniquely American. Cars got bigger and beefier. The average passenger car in 1950 used 627 gallons of fuel a year; by 1972 it was using 754: Fuel economy actually went down. And still, stations were offering anything they could to customers to grab more market share. Herb was giving Kleenex, stamps redeemable for panty hose, hula hoops, and cooking utensils. Self-serve caught on, but in a desultory way. By the late 1960s only 16 percent of stations in the country were self-serve.

And then came the 1973 oil crisis, when Arab oil-producing countries stopped shipping oil in retribution for U.S. support of Israel in the

Yom Kippur War. As the price of gasoline rose, the U.S. government implemented a rationing system that caused long lines at stations. Almost overnight the gas game changed and became something else. Drivers stopped thinking of stations as places to get pampered—the free maps, silverware, and stamps disappeared immediately. Instead, stations were places to get mad. People hated standing in line, and they started to hate going to the gas station. The cost of gasoline suddenly hurt, as did driving those big cars. People read the papers and saw major oil companies' profits more than double from 1972 to 1974. "Obscene profits," said politicians, and most people agreed.

The first oil crisis of 1973 was followed, like a one-two punch, by a second with the Iranian Revolution of 1979. Between 1978 and 1980 gas went from 63 cents a gallon (roughly $1.37 in 2000 dollars) to $1.19 ($2.20 in 2000 dollars). The oil companies, which had done their job of distributing gasoline and gifts by growing big and efficient, were now despised for those very same qualities. And little oil companies were not excluded from the anger. Herb remembers what happened when his company trucked in fuel from Utah. "San Jose had the highest price in the nation—79 cents—and in Salt Lake City there was excess gas. So we hauled it in. It was portrayed as gouging." Herb says his company did well, legally, during those years, but the atmosphere had changed forever.

Oil industry conspiracies are not something Herb favors, even though he spent years battling with the majors for market share. "I don't give them the credit to do that," he says disgustedly. "They don't have that much brains. And they'd get in trouble."

Self-serve gas suddenly looked like a good idea—who didn't want to save a few cents a gallon? By 1975 the number of self-serve stations had doubled, and by the 1980s full-serve was a relic. Tens of thousands of small mom-and-pop stations closed as the industry adapted to price-conscious consumers.

And people were driving less too. In 1980 the average passenger car drove 1,400 miles less than in 1972 and used 200 gallons less fuel a year. Cars became smaller and more sober.

In the new mood, the convenience store and the car adapted to each other. Self-serve begat the C-store as a way for the station owner to make profits. The C-store, with its ever bigger drink containers, begat the cup holder. And the cup holder became the new gas tank. Now a third of

U.S. car trips involve eating or drinking in the car. "In the U.S., the space between the two front seats is the most important place in the car. People are going to the gas station just to fill that spot," says consultant Craig Childress of the behavioral market research company Envirosell. He pauses before adding "Salad in a cup is a brilliant idea."

But when Herb looks around at the world he helped create, he suspects there are better ways to sell gas. "Convenience stores are a way of life, but I'm not completely sold on them," says Herb. "Something in my mind says that if a major is doing it we shouldn't do it. We've got to do something innovative. But I think I'm too damn old to be innovative."

**Regular Unleaded $2.45⁹⁄₁₀**
**Regular Unleaded $2.64⁹⁄₁₀**
**Regular Unleaded $2.57⁹⁄₁₀**
**Regular Unleaded $2.55⁹⁄₁₀**

By the summer of 2004, gas prices have risen. Floating high above the roadway, they are like a national stock ticker or football score. The price seems to offer an objective score of the country's economic prospects, but its impact is visceral. A glance at a price can be a comfort or a punch in the stomach.

Americans claim to hate high gas prices, but we sure love to talk about them. There is no safer conversational topic. It is better even than the weather. Who can't work themselves into a bipartisan froth over the cost of fuel?

Gas prices attained their superstar status during the energy crisis of the 1970s. Then, for the first time, a majority of Americans told pollsters that oil companies were "ripping off the public." In the years since, people's perceptions of the oil industry have been tied to gas price. When prices are low, as when they went below $1.00 a gallon in 1998, a majority feel the companies are doing a good job. But when prices rise, people punish the companies in opinion polls. By 2003 only 4 percent believed the oil companies were "generally honest and trustworthy."

"Belief in oil industry conspiracy is approaching cultural consensus," says political scientist Eric Smith, whose research has found that 85 percent of Californians believe that the industry is manipulating gas prices.

The suspicion cuts across social, political, and ideological lines, and it has big repercussions.

When gas prices are high, voters refuse to accept that they are the result of supply and demand. Politicians who work on timelines of six months to six years are expected to deliver to their constituents in short order. They try to harness voters' anger by attacking oil companies as "gougers," and ordering investigations. These antitrust investigations make good political theater, but they generally lead to nothing, actually reinforcing the sense that the public is powerless in the face of high prices. A 2006 Federal Trade Commission investigation of gouging is the latest in a long line of 200-page reports that have failed to find evidence of a conspiracy between oil companies to raise prices. At a time when demand is high and supply is tight, prices will rise.

But the myth of gasoline conspiracy overwhelms reason, particularly when pump prices and oil company profits are high. "Since the 1970s oil companies have come out as the bad guys," says Smith. "This is feeding into a story that's much simpler than supply and demand. It's more out of Hollywood than Economics 101."

The theater of punishing oil companies doesn't address the deeper issue of reducing demand or giving consumers more control over their gasoline destiny; it merely maintains the status quo. In April 2006, congressional Minority leader Nancy Pelosi stood in front of TV cameras to say: "The cost of corruption is so clear in the cost at the pump." She blamed Republicans for working in the interests of "big oil and the wealthy few." Her anger probably connected her with frustrated gas buyers across the country, but she knew that the real problem lay elsewhere. A few hours earlier she'd told a group of schoolchildren her real analysis: "There just hasn't been enough forward thinking to reduce our dependence on gasoline, and that is why the demand is high and therefore price is high."

Because the political melodrama around gas prices is largely a delaying tactic, oil companies, which work on timelines of ten to fifteen years, have figured out how to harness public anger for their own ends. They view anxiety over gas prices as an opportunity to change environmental laws. When prices are high, Smith's research has found, public opinion swings to favor offshore drilling, drilling in parklands, and other moves that are unthinkable to voters when prices are low. So when prices are high, oil companies withstand the attacks, which they know are temporary, while proposing regulations that will make their businesses more profitable in the long term. Voters, offered no other choices, support both drilling and investigations. And nothing changes, until the next time prices rise.

Consumers carry their rage to the gas pump. "For lot of people, buying gas is not a pleasant experience," says an engineer who's worked on gas pump design for thirty years. "They think the oil companies are ripping them off. I expect the oil companies have sleepless nights trying to figure out how to make buying gas pleasurable."

The engineer, whom I met at a trade fair standing in front of a gas pump he'd worked on, had a retro buzz cut and a happy snicker in his voice. People are intimidated by big gas pumps, he said, but they have warm feelings for ATMs. So this pump is designed to look like a friendly ATM. He gestured. The pump was small, sleek, and didn't have any greasy surfaces. It had a soft rounded belly with an ATM on the front. The pump looked playful, unobjectionable, maybe even meek. "The ultimate goal is to make buying gas a pleasurable experience," said the engineer with a wry smile, "so I don't feel like I'm giving the oil company my money." He broke out into a chuckle, repeated the line about "ripping off the public," and added, "and we make the machine that helps them do it!" At that point his boss stepped in and said he'd answer my questions from now on.

Since that encounter, I've gained a new appreciation for how neatly gas pumps embrace contradictions both political and mechanical. Charles Keane is an industrial designer who worked on the first popular self-serve pumps designed in the 1980s. "It was mayhem and murder," he remembers, to train customers to use the machines to pump their own gas. Designers now, he says, are just putting the "icing on the cake because the cake [self-serve] was digested long ago."

The icing, it turns out, is trying to make people forget what they're buying by "decommodifying" gasoline. Keane's design group tried to rethink the gas station experience, with its greasy smell, its ugly pumps, its sense of danger and wasted time. They conducted focus groups and watched hundreds of people buy gas, including a woman in a station who locked her infant and her keys in the car while filling up. "That certainly indicates that apprehension is the operating paradigm in play," says Keane, "but it was a nice station and they lent her a car to go home and get her keys."

Keane's crew ended up designing a tall thin pump, which made filling SUVs easy, shed lots of light, allowed customers to make eye contact with the attendant, and could offer Internet access. And one more thing: It resembled a tree. "Most sites are ugly," says Keane. "We wanted to incorporate natural forms. The more you reconfigure the [gas-buying] experience towards the positive, they might feel better about themselves and go for a fill-up."

The pump does a neat job of selling gas without focusing on the gas. But here again, it has to deal with a contradiction. If asked, people will say they shop for gas based on price. For gas that is 3 cents cheaper, 40 percent of us say we'll switch stations; for 8 cents, all of us say we'll switch. Yet when it comes to gasoline, marketers say consumers are unreliable narrators: We lie. "The vast majority are puritanical," says Keane. "They will never admit they shop for convenience, entertainment, and looks."

### Regular Unleaded $2.55%10

It's the summer of 2005—two years since I first hung out in Twin Peaks. I've come back periodically, but this time the station looks bleak. Where the poster of the snowman drinking the cola hung there is a plain white sign with red lettering: HELP WANTED TAKING APPLICATIONS. A hearse is parked at one of the pumps.

B. J. is nowhere to be found. Talking to the young Nepali clerk is a struggle. The shelves are a war zone between five flavors of corn nuts, four flavors of Skittles, five flavors of Cheetos and eight of Doritos. The candy wizards are gone. Energy and power bars occupy a whole shelf. Their names sound like gasoline additives: Protein Plus Carb Select Chocolate Peanut Butter." In the drink vault there are energy drinks named Rock Star next to Slenderize and Meal Replacement Drink.

Beneath its dull facade and bizarre array of products, the C-store gas station is a perfectly tuned reflection of my irreconcilable desires and peccadilloes. "Neotraditionalists," according to a ConocoPhillips study, want ATMs and convenience but with service delivered "the way it used to be." Gimme a rocket ship and a scullery maid. Or give me oil from a village in Ecuador at a price low enough that I can treat myself to pure spring water. Or a drink that will make me thin. Gas pumps shaped like trees? Sure. At the gas station, there are no limits.

Michael Gharib comes in, again in a perfectly pressed pink shirt, wearing rimless glasses. He talks briefly with the clerk, who grabs a pole and some new numbers and heads out to change the sign. Gharib is, as always, upbeat, but he is not optimistic.

"I have to change prices two times a day, maybe more, as the market has become more volatile," he says apologetically as we watch the clerk post the numbers.

I ask how the ATM worked out. It didn't. It took up too much space and didn't bring in enough profit. Meanwhile, the industry has only gotten tougher.

"Before, we were inverted two times a year. Now it's half the year that our prices are higher than the branded. The refineries are running at capacity, which makes the market volatile. Even if there's a rumor of something in a refinery the prices go up like a rocket and then trickle back downwards. As we speak the branded prices are 7 cents below mine—these guys are selling gas below what I buy it for." He shrugs. He says it would almost be worth it for him to buy gas from his competitors and resell it.

He takes me back to his office, where the mail is still sitting between the minarets of the Taj Majal. Now, he says, he's making a few hundred dollars a month by putting ads on top of his gas pumps. He pulls out next month's ads: Mini Cooper cars, MasterCard, and long-distance calls in Farsi.

"It's just another thing I need to keep track of," he says, rolling his eyes. "I was a gas station, not an advertising firm. It's an evolving business. You really have to stay on top of it, and it's very difficult. I am closer and closer to closing."

B. J., he says, left to start his own convenience store with his family. "That was tremendous," he says. He looks exhausted.

I go out to the yard. The sign has been changed. Now it reads:

**Regular Unleaded $2.59⁹⁄₁₀**

What is the deal with the 9/10?

Gharib laughs, looking relaxed for the first time. "It's the invisible penny," he says. "It's a fixture. It's the built-in penny that nobody sees. We're trying to get everything we can." Every gallon of gasoline sold in the U.S. includes the invisible penny. That's 140 billion gallons of gas a year. The invisible penny is worth $1.26 billion a year, and it speaks to what shoestring operations our corner gas stations are.

The hidden penny on the sign is the first in a long line of hidden pennies (and a lot of dollars) that lurk in the external costs of gasoline—from air pollution and greenhouse gases, to traffic congestion and tax breaks for oil companies, to taxpayer-subsidized oil investment in developing countries, to the costs of the U.S. military presence in the Persian Gulf. The hidden penny is a placeholder for the unquantifiable costs of oil extraction—among them environmental damage, poverty, and human-rights violations in oil-producing countries. Hidden pennies determine how our government makes policies to conserve fuel—because we evaluate fuel's price rather than its intrinsic value or replacement cost—and hidden pennies mislead us. No matter how scrupulously we shop for gas, we have no idea what the stuff is really costing us. Eventually, we may pay a high price for our ignorance.

Another revelation in the hidden penny is that most Americans literally have no idea how much money they spend on gasoline. Studies have shown that few of us, myself included, can remember the cost of our last fill-up, or know how much it'll cost us to drive to the mall and back. And of course, no one adds the cost of the soft-drink or bag of chips to the price we paid to fill the tank.

If you ask people whether they're being hurt by high gas prices, they will answer emphatically. Between 2003 and 2005, the price of gasoline increased by an average of 77 cents a gallon—really a huge jump. But when the Harris poll asked people to estimate how much more they were spending per month on gas than in previous years, answers for the difference between 2003 and 2005 added up to an increase of $3 per gallon. They overestimated gas price increases by nearly four times. Were they lying? Exaggerating? Did their anger overwhelm their mathematical ability? Perhaps all of the above.

Even though we talk a lot about gas prices, most of us don't spend much on the stuff. The average American household spends only 3 or 4 percent of its income on gasoline. In 2005 the Automobile Club of America (AAA) estimated that the average car owner spent 8.2 cents per mile on gasoline and another 60 cents per mile on maintenance, insurance, depreciation, and finance charges, among other things. So it's not particularly amazing that in 2004, a tiny 4 percent of car buyers said fuel economy was their most important criterion when buying a car, according to surveys of 90,000 buyers by Maritz Inc. Cup holders were a higher priority than fuel economy.

Could it be that years of buying gas as an idea has made it difficult for us to think of it as an expense? "I earn more money by presenting myself as successful than I could save driving an economy car," a banker with a luxury SUV told University of California researchers Ken Kurani and Thomas Turrantine. He is hardly alone. Sometime during the last ninety-eight years at the gas pump, gasoline became a component of our self-esteem, part of the way we package ourselves instead of a commodity. The researchers charitably concluded that consumers "engage in a type of limited economic rationality" about fuel.[6]

Economists have tried to figure out how high prices would have to go before people cut back on the amount of gas they used. One study found that if prices stayed around $4 for an entire year, gas consumption would only fall 5 percent. It's too hard for people to cut back quickly. Drivers told researchers that they wanted to stop driving where they had to—work, school, errands—and didn't want to stop driving to the places they enjoyed going to. After Hurricane Katrina sent gasoline prices over $3 a gallon in the fall of 2005, credit cards, which had been fostered by the oil companies decades ago, came to drivers' rescue as 85 percent of gas buyers used credit cards to fill their tanks. (When prices were lower, credit card use was closer to 55 percent.)

So if there's no real price limit on how much we spend, and no larger economic rationality to our fuel use, what's to fence us in? Not much. Even as people complained of rising gasoline prices between 2000 and 2006, national consumption of gasoline continued to rise. This "limited economic rationality" about gasoline seems like a personality quirk, but it has profound effects on other parts of the petroleum universe. From the truck drivers, to refiners, to the international oil traders, to Venezuela. In order to see them, I'll have to leave Twin Peaks and start venturing up the supply chain.

---

[6] Some out-of-state gamblers said they'd still drive to casinos in Las Vegas even if the price of gasoline were $10 per gallon, according to a 2005 study by MRC Group. I admire their honesty, though I don't want to imagine the catastrophe that would lead to $10 gas or the catastrophe that would result. On second thought, maybe a casino is exactly where you want to be when gas is $10 per gallon.

Even if we're not rational about gasoline, you might think that traffic, which is getting worse as more people get on the road, would stop us. People who live in crowded areas now spend sixty-one hours a year stuck in traffic, about three times what they did in 1983. But studies show that people love being alone in their cars even more than, say, being at work or at home. People actually prefer a half-hour commute to a shorter drive.

"There seems to be no upward limit to the amount of time Americans will spend in their cars," says auto industry anthropologist Steve Barnett. "When you drive, the car becomes who you are. You can be anything, do anything. You construct your own universe. You can transform your personality." For now, cheap gasoline is the great enabler of this imaginary universe. Barnett continues: "Why is it that there are a consistent number of arrests year on year for driving nude?"

# 2 | DISTRIBUTION  *WAITING IN THE TRAFFIC JAM*

"Sitting in traffic makes me mad," says Roger, without sounding mad in the least. Most days he spends twelve hours driving a shiny tanker truck to deliver gasoline or diesel from the pipeline to stations and customers around San Jose. A lean man in his late forties, Roger stands next to the tanker slightly hunched, as if he's still hunkering over the steering wheel. It's a high-stress job—there's the flammable cargo, the anxious dispatcher on his walkie-talkie, the even more anxious gas station owners who fear running out of fuel on days like this, and the long lines at the "rack," or pipeline terminal where the tankers pick up fuel from the pipelines. Not to mention traffic, the California Highway Patrol, breakdowns. Roger absorbs it with jittery resignation. "It's all attitude." In his mirrored wraparound sunglasses, he slightly resembles the truck he drives.

Tanker trucks are beautiful things on the highway, their silver sides reflecting the streams of traffic around them, as if to prove that none of us would be moving at all if it weren't for their presence. There's something Victorian about the silver convex oval on the rear of the tanker—it offers up a pastoral reflection of the freeway landscape around us while primly shielding its own contents. But if you look carefully at the oval you'll see a smallish diamond-shaped sign with a number on it. The number 1203 indicates gasoline; 1993 is diesel.[1]

---

[1] Diesel fuel is heavier and more "oily" feeling than gasoline, because the hydrocarbon chains that make up its molecules are longer. Diesel burns at a higher temperature than gasoline and evaporates more slowly, but diesel engines are more efficient than gasoline ones. Diesel can be made from things other than petroleum, including vegetable oil, turkey parts, natural gas, and coal.

Roger wears a crisp company button-down shirt with pressed long denim shorts, socks, and sneakers, achieving a look both formal and cantankerous. He waves vaguely at the shorts. "Safety hazard," he says sarcastically. Earlier this morning, Roger made his first trip of the day to the rack to pick up a load of diesel. The manager saw his shorts, told him they were out of line, and demanded he wear long pants. Roger disagrees. He says he's been wearing shorts to deliver gas for twenty-three years, since the days when San Jose was still a bunch of prune orchards. "If I spill gas on my legs I can hose off. But with Levi's, if you hose off and dry they're still a torch." Neither of those options sounds good to me, actually, but Roger lives with combustibles in a way most of us don't. "I think you're safer with shorts," he says. Tanker driving is a risky job with a burdensome (and necessary) number of rules, and yet it attracts a bit of an outlaw spirit.

Tanker trucks are the final and most visible part of an enormous octopus of pipelines, rail lines, barges, and tanker ships delivering gasoline, diesel, and other petroleum products from the oil fields, to refineries, to gas stations. Daily, men like Roger make 50,000 deliveries of fuel from pipeline depots to gas stations. Load by load, this unseen network of drivers, dispatchers, pipeline engineers, and traders keep the rest of us running. Today Roger will deliver 8,000 gallons of fuel.

Roger waves me in the truck and I scramble up the foothold and heave myself onto the bench seat. The truck rides high, stiff, and jouncy. Roger hauls fuel for Coast Oil, a Northern California jobber, or fuel hauler and wholesaler, that has been in business since 1935. Roger runs through the gears as we bounce out of Coast's yard toward the freeway. Soon we're in the slow lane. It's the Friday before the Fourth of July weekend, already a cooker, and the traffic is going in fits and starts, cars aggressively juggling lanes. Near the steering wheel hangs a tiny truck-size pinup calendar with Haley, Miss July, wearing a tasteful pair of flag shorts, a red baseball hat, and nothing else. I ask Roger if he worries about crashing.

"It don't bother me," says Roger. "Not too many of these things get in a wreck and blow up. It's usually a fender bender." I think about that for a moment. What is a "fender bender" in a tanker truck? According to the Department of Transportation, there have been between 6,000 and 7,000 fuel tanker accidents a year since 2000. Only about 1 in 1,000 of those accidents have been considered "major," though they've killed forty-nine people over the last five years.

But when a tanker does crash, it's horrific. Roger drives a truck called a bobtail, which can hold 4,200 gallons of gasoline, which weighs about 13 tons. Mixed with air, and set on fire, the gasoline releases the same amount of energy as 194 tons of TNT. Most of the tankers you see on the road haul a double trailer, which means they hold twice that amount. When a double trailer accidentally overturned near the Pentagon in December 2004, neighbors thought it was a terrorist bombing. Flames a hundred feet in the air melted a streetlight; the truck's cab was reduced to a piece of char; and burning fuel ran into the sewer, where the heat blew off the manhole covers.

Roger never, ever drives faster than fifty miles an hour—even when he's in a horrendous hurry. Caught speeding once, he'll have to pay a $500 fine. The second time a fine and six-month suspension. The third strike is out. His employer tracks his speed with a satellite system as well as an old-fashioned setup involving a needle and a disc of paper to record accelerations. After two decades of driving, the speedometer has become embedded in Roger. He says he's incapable of going faster, even when he's out with his family. "My kids yell at me," he says.

We're on the freeway now, heading toward the rack to pick up fuel, moving eighteen miles per hour, cars pushing in from all sides. On the edge of the traffic the mountains look flat, smashed down by the greenish brown haze.

Traffic jams are a symptom of prosperity. In the late 1990s, people enjoyed the growing economy by buying homes farther from where they worked. To compensate for the increased commute of around 1,200 miles a year, they bought bigger cars, adding an extra 65 gallons of fuel consumption per year to their household bill. More cars driving longer distances means more traffic jams, which themselves eat more fuel—2.3 billion gallons of gas a year, according to a study by the Texas Transportation Institute. And as bigger trucks and cars clogged onto freeways between 1993 and 2003, they required 1.6 percent more fuel every year. And in a jam Roger and the gas can't get through.

Roger's walkie-talkie crackles. A new driver lost his clutch. It's going to take a wrecker an hour to get there through the gridlock. "Get out and push," Roger snorts before explaining how to start the truck without the clutch to get it off the highway. "He didn't grow up on a farm," he scoffs.

The hoarse voice of Chris, Coast's dispatcher, comes on the speaker. She wants to know if Roger can work a twelve-hour shift again. "Prices are up all over the place. I got everybody jumping and begging and we're out a truck and the racks are backed up," she says. He agrees to stay, with an amiable crankiness, the it's-all-attitude thing. When I ask if he gets tired of the long hours, he looks at me curiously, as if I'm a small child, a Chihuahua, or a rich heiress. "Seems like everyone I know puts in a lot of hours." He started work at 5:00 A.M.; it's 9:58 now.

The rack is really backed up. Nine silver trucks, each pulling an extra tanker trailer, snake out toward the road. An unimpressive dusty concrete yard, the depot contains thick clumps of pipes and hoses sticking into the air like small trees. The drivers stand in a patch of shade grumbling. Roger exchanges wisecracks, drives around the line, and pulls his truck under the tree of pipes that dispenses unbranded fuel.

I've always thought of gasoline as something that comes out of a small, human-size pump, surrounded by colors and words signifying a brand—even if it's something generic like Twin Peaks. But at the rack, the few brand placards on top of the dispensers are sober and perfunctory. Here fuel is fuel, and it's a bit of a shock. Huge white tanks of the stuff sit out in plain sight. The dispensers are bare-bones assemblies of exposed pipe, wiring, and control equipment. Gasoline at this rack comes from several different refineries, but it all travels in the same pipes and is interchangeable until it reaches the dispenser. Whatever the difference is between brands, it is only this deep.

The rack is where gasoline becomes the stuff we know and use. One hose adds ethanol into the gasoline so that it meets federal clean air standards. And while high-octane gasoline and low-octane gas travel separately through the pipeline, at the dispenser they're mixed in equal parts to make midgrade octane. The difference between brands comes into being here too, because this is where the proprietary detergent package squirts into the fuel, making it one brand or another.

Roger calls the dispatcher to be sure he's buying the right fuel. Diesel, she says. He takes a plastic card into a small shack, swipes it, keys in a stream of numbers. When the fuel enters his truck it becomes Coast's property—the swipe represents thousands of dollars. By purchasing the fuel, he has also prepaid state and federal fuel taxes.

Taxes are added to the price of gas at the rack so that the haulers can pass them on to the stations, which then pass them on to drivers. Fuel taxes were conceived by states in 1918 as a way to get drivers to pay for roads. Back then, states lacked the technology to charge drivers for the miles they drove, so they charged by the gallons of gas they used. The taxes were put on the gas at the rack because it was easier for states to collect them from a few central points. Off-road vehicles, such as tractors, farm trucks, and forklifts, are allowed to buy untaxed fuel, because they don't use the roads.

Taxing gasoline gives the state a perverse interest in citizens using more gas. In 2004 state governments collected $35 billion in gas taxes, and in some states some of that money was diverted to state projects other than roads. "Do we serve ourselves well when the construction and maintenance of our highways depends on increased fuel consumption?" asks Dr. Martin Wachs of UC Berkeley's Institute of Transportation Studies. As cars get more efficient, the roads get less care. "We've outlived the gas tax, but there's little incentive to change it. We need to do something so the government isn't in the place of a drug pusher."[2]

To fill the tanker, Roger hooks up a series of hoses, sealing the nozzle into the tank and attaching a cable to ground the vehicle. Today he's delivering diesel to a wrecking yard and a private company's fire station—two places that don't have to pay tax on the diesel. To prevent people from buying off-road fuel and using it to drive on the highway, the state dyes off-road diesel pink, so police can see what's in the tank. Standing in the rack, under the connector hose, I can look up in one of the chambers and see red dye squirting out into the diesel like a squid spraying ink. Roger says he has to be careful he clears the red diesel out of his tank before he fills it with regular, taxed diesel. If he leaves any red diesel in the tank and then adds the regular, he could end up with pink diesel, which is a problem, from the point of view of the police, who stop fuel trucks at random.

---

[2] Are high gas taxes a way to motivate people to use less of the stuff? They appear to offer an answer, because you'd assume that making gasoline more expensive would make people use less of it. But Dr. David L. Greene, an energy economist with Oak Ridge National Labs, questions whether gas buyers can be expected to operate in a "textbook economically rational way" around fuel economy. Taxes, he says, are an important way to signal to consumers the importance of using less gas, make it easier for car makers to sell more efficient vehicles, and prevent people from driving more the moment they get a more efficient car, but changing demand will require more. As proof he points to Europe, where gas costs two to three times as much as in the United States, but governments have still resorted to setting strict fuel efficiency goals for auto manufacturers.

Fifteen minutes later we're back on the freeway, and the suspension is no longer bouncy but rocking ponderously—heavy fuel sloshes slowly. Roger says this truck has soft suspension, like a Cadillac. "I'm a spoiled baby," he says.

The road is clearer now, and we rock along in the slow lane while Roger fumes a bit about government regulations, calling them "job security for the government." If he overloads the tanker, he can be sure the highway patrol will wave him over into a weigh station because they know this is exactly the sort of day when truckers could be inclined to cheat.[3]

Hot aggravating days are also the times when fuel delivery drivers are likely to make mistakes, such as dumping the wrong fuel in the wrong underground tanks at gas stations. Roger's truck is outfitted with an extra pump that allows him to pump gas out of an underground tank and into his truck. So it's usually Roger to the rescue. "Everybody makes mistakes," he says with a shrug.

And then there are the Environmental Protection Agency regulations about spills, so many rules he says it's hard to keep up. He carries towels to clean up dribbles, and is supposed to report any spill of more than a gallon. "Talk about the way things changed," he says. "The old guys—if they had an extra thousand gallons that wouldn't fit into a gas station tank, they'd just dump it in the creek." The walkie-talkie is snapping again.

He tosses me his guide to hazardous materials. I open it to a random page and read, "Insecticide travels downwind 8.3 km at night." "I wouldn't drive if I read that book," he says.

What would you do if you had a spill? I ask him. "Run," he jokes.

He starts to muse on the possible dangers underlying the job. "I've heard it's dangerous and it's gonna give us cancer, but I deal with it every day. I can't dwell on these things." He's quiet for a minute. I can feel the

---

[3] Petroleum chemistry is one reason the police bust overloaded trucks on hot days when there are traffic jams. Roads made of asphalt, a by-product of crude oil, behave like chewing gum; because their viscosity changes in relationship to heat and frequency. If you stretch chewing gum quickly or in the cold, it's stiff and will snap. But if you pull gum slowly or in the heat, it will stretch and stretch and stretch. On a cool day, asphalt stays stiff and trucks do little damage to roads, but on a hot day, particularly when very heavy trucks are moving very slowly in traffic jams, the asphalt begins to stretch and smoosh underneath their tires, forming ruts.

tanker sloshing. "It seems like a lot of people in the industry get cancer, but you always think it can't happen to you."

It's a stoic industry. Later, back at Coast's yard, an enthusiastic employee leads me back among the storage tanks where he mixes lubricating oils. He mentions some of the risky chemicals he used to handle before they were outlawed. Then he starts talking about the cats that live near the tanks. "Third generation with six toes on every foot. Weirdest thing. Like they're walking with snow skis on. Puddle drinkers." I ask to meet some of these creatures, but a tour of the tank farm comes up catless. I was also unable to verify his claim that the pigeons in the yard don't have blood in their veins.

I can't find any studies showing that gasoline creates six-toed cats. Gasoline consists of 150 chemicals, and according to the Agency for Toxic Substances and Disease Registry, rats exposed to high levels of vapors continuously for two years developed cancer, not extra toes. Those rats got a bum assignment. While the agency says there's no evidence that gasoline exposure causes cancer in humans, benzene, which makes up about 1 percent of gasoline, is a known carcinogen, causing leukemia even at very low exposure.

I called Dr. Dave Verma, who's been studying benzene and truck drivers for more than twenty years at McMaster University, to see just how unhealthful Roger's job is. I was hoping for a neat number reflecting relative risk. But there are far too many variables for that. "Driving a petrol truck is a higher risk than driving a vegetable truck," Verma conceded, "because the tanker drivers have added exposure to the petroleum." All truck drivers are at risk because they spend most of their working time inhaling diesel exhaust from the trucks around them in traffic. But the fumes a tanker driver inhales while loading and unloading his truck depends on his specific job. Verma's research suggests that Roger, who makes lots of little deliveries, probably inhales a lot more than his coworkers who do bigger deliveries to gas stations.

A big puzzle, according to Verma, is figuring out how many hours the drivers are working. The normal exposure limits are calibrated for eight-hour days. "One bulk driver was going fourteen or fifteen hours—even though they're not supposed to," he said, somewhat exasperated. These hardworking drivers were messing up Verma's calculations, not only because they were increasing their exposure, but also because they were decreasing their downtime between shifts. The gas is getting through, but the drivers are paying some unknown price in increased risk.

Roger pulls off the highway into an industrial zone and drives into a wrecking yard. There are stacks of gutted cars here, and piles of tires and batteries. A forklift zips by holding a Chevy van ten feet in the air. $2000 OBO. AS IS NO GUARANTEES is written on the window. Definitely no guarantees. "It's the end of the road," I say. "Huh," Roger replies, "or maybe it's just the beginning." I wasn't about to pester Roger about his theory of life, but if I had to guess, it would be that the world is a continuous loop of tankers going an even fifty miles an hour, always picking up and dropping off, loading and unloading, coming and going, twenty-four hours a day.

Soon we're back on the highway. It's almost noon and we're moving toward our destination at an even rate of speed, with no bad news and none particularly good, either. "I do my little job," says Roger. "Nobody messes with me and there's no politics." On a perfect day, the fuel will arrive so smoothly that no one will even notice.

**Chris, Roger's dispatcher,** works from a utilitarian room on the side of Coast's yard, where no truck entering or exiting can escape her appraising glance. A bottle labeled "Chill Pills" sits on her computer and a sign reading THE MORE YOU COMPLAIN THE LONGER GOD MAKES YOU LIVE hangs behind her. All day long she alternately coos and growls as she dispatches gasoline across Northern California.

On her bulletin board there's a pinup calendar of tanker trucks that hasn't been changed since last month. Miss June is a Peterbuilt 385 truck posing coyly in front of a tank farm.

Chris was a delivery truck driver for eight years before moving into dispatching. She has long legs and long blond hair and is a classic California beauty, but she's adopted a bitingly tough shell, not at all diminished by her smudgeproof pink lipstick and oversized T-shirt with a photo of two wolves kissing. When she scowls, which is often, her black eyebrows form angry little tweezers over the bridge of her nose. Regardless of whether she's cooing at customers or growling at the truckers, she keeps her molars clenched. She has two phones, a walkie-talkie, a fax, a price feed, e-mail, and a pager.

Last month she bought, sold, and dispatched 886 loads of gas—8 million gallons.

To deliver fuel to the one hundred independent filling stations that Coast serves before any run out of gas, she needs to be planning deliveries of four different types of fuel all day and all night, allowing for traffic jams, backed-up racks, and drivers who sleep too late. But that's not all. She also needs to be buying that gasoline from the cheapest supplier at the moment—there are four selling unbranded gas in this area at different locations. The buying and routing of all of this fuel is a logistical challenge, but it's made more difficult by the fact that Coast's profit margin hovers around .25 cent per gallon.

"A good dispatcher has a bit of larceny in 'em," says Chris's boss, Mark Mitchell. "They stay calm when everything is going nuts. You kind of have to love being crazy every day."

This morning was a case in point. When part of one of the seven local refineries went down, wholesale prices zoomed up 6 cents in less than ten minutes—a huge rise in an industry where prices are calculated to the hundredth of a cent.

On the phones, Coast's customers panicked, trying to buy gas before the price rose further. As the phones started to ring, Chris's price feed froze, so she didn't know which way the prices were going. And in the midst of the bedlam, Chris had to decide where two drivers would buy gas. She told them to buy gas that was $1.0695 a gallon, assuming that the price had climbed. She made the wrong call because the price had already fallen to $1.0395. Off by 3 cents. The company lost $492 on the two loads. With thin profit margins, it'll take ten loads of 8,800 gallons to make up the loss. "It killed me," says Chris when Mark comes to talk with her. She uses a third voice, which is just tired.

Mark is puzzled by the price rise. He heard that the refiner only lost 20 percent of its production, which means the region's supply is out by only 4 percent, and yet the wholesale price rose by more than 10 percent. This is part of the volatility that started in California's gasoline market in 1999 and has gotten worse year by year. With the close of several refineries in the late 1990s, the state began consuming more gasoline than it produces. "With a system that fragile, you're carrying billions of gallons one load at a time," says Mark.

When seesawing prices first hit in 1999, it seemed like a local thing, brought on by California's isolated network and "boutique," low-polluting gasoline. But by 2005 volatility was appearing in the rest of the country too. "Clearly demand has exceeded supply for the nation," says Mary Welge, of OPIS, a company that tracks wholesale gas prices nationally. "Demand is at record highs for the country. There's a daily fear of a refinery going down and parts of the country being in a hand-to-mouth situation." All of the pipelines are booked with deliveries now. "All indicators for the future point to high demand and short supply," says Welge.

Ten minutes of Chris's afternoon: Phone rings (musical singsong): "Hel-lo, this is Chris, I'll try. If it's loaded today it's today's price and if it's loaded after midnight it's tomorrow's." The walkie-talkie gets the growl. "What's the earliest?" The fax rings and an order comes in. A man appears at the window trying to get a truckload of lubricating oil. Chris orders a sandwich: three slices of cheese and nothing else. Nothing else! Two more calls. The walkie-talkie gets clipped on the belt. Chris organizes a sheaf of thirty invoices as if she's playing an elaborate game of solitaire, muttering to herself as she schedules the deliveries through the night and until eight the next morning. Phone (coo): "Yep. I'll try. I've got a truck down." Walkie-talkie (growl): "Without that truck I'm dead."

By 1:11, two of the four racks have shut down and the other two have lines of three hours instead of the usual forty-five minutes. The price is going up. Gas station owners are panicking. "I've stopped trying to guess the market," Chris says. "As soon as somebody hears of a problem at a refinery, the price starts to climb." When prices climb, there is no relief. Many refineries making California's unique blend of high-air-quality gasoline are at least twenty days away by tanker—in Trinidad, Newfoundland, and Finland. Mark sends her a text message that he thinks the price may start to fall. "I gave up," Chris says. "I had too many wrong guesses."

Chris hasn't had a day off, or a night when she wasn't on call, for more than a month, when she was sick for one day. "Wherever I go 90 percent of the time that briefcase goes with me," she says, pointing to a heavy case filled with charts that allow her to convert wholesale prices to the sale price for the dealers. Inside the case is a smaller folder containing smaller charts. "I take that if I go to eat," she says.

If we bought gasoline the way people buy widgets in Economics 101, we'd stop consuming gasoline when prices spike, reducing demand and allowing Chris to get some sleep at night. But that's not what's happening. Huge price jumps don't shock people into using less gas. One study based on a 2003 supply shortage in Phoenix caused by a broken pipeline suggests that gas prices will have to increase by 150 percent to get people to decrease the amount they use by 30 percent in the short term. Drivers either can't cut back or they don't know how to, so they get angry, and people like Roger have to deal with them. After that pipeline break caused gasoline shortages, high gas prices, and anger toward oil companies, Arizona's governor set up an 800 number so gasoline tanker drivers could call police escorts to protect them at gas stations.

Another phone call: Chris growls at a trucker who insulted a gas station owner. "Look. The competition's really tough up there and we need to kiss butt. It's not much but this industry is all about counting pennies."

As the afternoon winds on, the price falls a bit, and the pressure drops. Chris organizes her loads, and organizes them again. Coo. Growl. Coo. Growl. The office has a limp, all-jittered-out feeling. I ask her if she likes her job.

"Once in a while I miss driving," she says without looking up from her work. "The only thing I don't miss is the traffic. There are a bunch of idiots out there. You get into a station and they'll drive all over your cones [the orange cones tanker drivers put out to prevent people from driving over their hoses]. Their excuse is that they don't even see it." A driver comes in, spies Chris's partially eaten sandwich, and asks if he can have it. He leaves stuffing it into his mouth.


**At the end** of a long day Mark Mitchell returns to his utilitarian office in the back of Coast's building. Mark is in his late forties, retaining a boyish look with his halo of curly ringlets. Pictures of his days as a tennis player at Stanford hang on his office walls. One shows him suspended in midair, feet off the ground. Now his quick-twitch reflexes are confined to an office that's almost studiously bland. OPPORTUNITY: THERE'S AN IS-LAND OF OPPORTUNITY IN EVERY DIFFICULTY says the poster behind him. The magazine *Lubes 'n Greases* leans against *Fast Company* on the file cabinet.

Being a jobber is a trade that started at the turn of the last century, and as mergers have consolidated the oil industry, it has had to change fast. "Darwin wasn't wrong," Mark says. "This is the third wave of small jobbers closing. Most of the guys haven't changed with the times. And a lot are dead and don't know it." He speaks in short bursts, as if typing numbers into an adding machine. "The cash flow keeps going even when you've been losing money for a long time, without realizing it, until one day the banks stop your credit. Costs rise. Exposure rises. Margins are decreasing. You need to have volume to be efficient."

When Mark and his business partner, Roy Youngquist, bought Coast, they promised the employees they'd keep the company going until 2035, when Mark will be eighty and the company will be one hundred. Roy had already been with the company for thirty-two years anyway. Mark jokingly calls the business an albatross.

Keeping Coast alive is not going to be easy—this is an intensely competitive, high-stakes, low-rewards business—but Mark is counting on it being fun. He comes to work before six in the morning, leaves after seven at night, and keeps his cell phone on except at church and the movies. "We installed a shower last year," he says. "You can't compete if you don't run twenty-four hours a day. That's what keeps the whole country driving."

When Mark started at Coast, the company delivered 12 million gallons of gasoline a year; now it delivers 8 million a month. He wants to sell 20 million gallons a month. But he'll need more trucks, more dispatchers, more customers, and more gasoline to sell them. He'll have to finance ever larger gasoline purchases, while insulating the company against the possibility of customers who can't pay their bills. "The truth is: How long is the small garage or independent station going to stay in business?" he says. "I don't know where it will all end. It's kind of scary if you can't grow big."

**The Energy Information Agency** estimates that, by 2025, Americans will use 37 percent more petroleum than we did in 2004—rising at a steady rate of 1.5 percent per year. That's a 37 percent increase in the amount of gasoline that will need to be shipped in tankers, marine terminals, barges, pipelines, and tanker trucks. At the moment, pipelines (which ship half of the gasoline) are running at 90 to 96 percent capacity. Oil moves very slowly in pipelines—at around three to eight miles per hour—and it's not possible to increase that speed. "Obviously there's no way we're going to get that kind of increase into the pipelines," says Bob Reynolds of the petroleum transit consultancy Downstream Alternatives. Waterways are already backed up. Reynolds says that barge shipments are congested and losing whole days just waiting to get through antiquated locks on the inland waterways. And because the Jones act requires tankers running between U.S. ports to be U.S.-owned, the number of tankers to take up the slack is limited. A traffic jam is developing in the fuel supply chain. "From the terminal to the pipeline to the refiners," says Reynolds, "same game all the way back."

So how will these extra gallons of gasoline get distributed? Reynolds imagines that some of the increased fuel will be shuttled on tankers from one place to another. But eventually the 161,000 miles of fuel pipelines in the United States will need some additions. (Only 8,000 of those miles of pipeline have been built within the last twenty years.) Reynolds estimates that the country will need another 10,000 miles of pipeline to handle the difference, at a cost of $1 trillion or so.

And there are other indirect costs to using more fuel. "The more oil being transported and used, the more opportunities there are for spills to occur," says Dagmar Schmidt Etkin, a consultant who's worked on oil spill issues for nearly twenty years. Before I talked with Etkin, my image of an oil spill was dominated by the *Exxon Valdez,* which spilled 11 million gallons of crude oil into Prince William Sound. I did not connect oil spills with the fuel in my car, but thought of them as the fault of oil companies.[4] Stricter liability laws and increasing costs for spill clean-up and damage have motivated the industry to work to reduce the number and size of spills, even though more oil is being imported, shipped, and used. "As long as we're dependent on petroleum, there will always be spillage to some degree," she says. "We all spilled oil in Prince William Sound. It's part of our lifestyle, something we all need to reckon with."

Spills are one of the costs of oil consumption that don't appear at the pump. Etkin's data shows that 120 million gallons of oil were spilled

---

[4] Drivers and boaters spill more oil every year than the Valdez. Leaking oil from U.S. cars and trucks and two-stroke engines adds nearly 19 million gallons of oil to waterways and the sea every year, according to a 2002 report by the National Academy of Science.

in inland waters between 1985 and 2003. From that she calculates that between 1980 and 2003, pipelines spilled 27 gallons of oil for every billion "ton miles" of oil they transported, while barges and tankers spilled around 15 gallons and trucks spilled 37 gallons. (A ton of oil is 294 gallons. If you ship a ton of oil for one mile you have one ton mile.) Right now the United States ships about 900 billion ton miles of oil and oil products per year.

Spills can be horrific. A 1999 pipeline break in Bellingham, Washington, released nearly 250,000 gallons of gasoline into a creek, formed a fireball, and killed two ten-year-old boys and a fisherman in a local park, decimating the surrounding area and poisoning the waterway. A report on pipeline safety by the Government Accountability Office notes that while pipeline spills in general fell between 1994 and 2002, the rate of serious spills—with deaths, injuries, or property damage—stayed constant at .15 per billion ton miles, and recommended more regulation and tougher enforcement of the pipeline industry.

**In the afternoon** Mark calls some refinery managers to find out what their stocks look like. He accepts delivery of a load of fuel from a refinery that he doesn't particularly want because he knows that if he helps the refinery now, it'll help him later. He calls Chris. "Let's get rid of everything [in Coast's tanks] over the weekend. It's gonna crash on Monday." He makes a quick decision to buy more fuel from a refiner even though a wholesaler has a slightly better price. "We never compete with our suppliers," he says. "Jobbers are notorious for leaking—buying discount gas at truck stops and reselling it to wholesalers."

Later he sets prices for the stations Coast supplies based on the distance the trucks have to travel. Some of the company's clients, like Twin Peaks, have a long relationship and years of good credit. They get a slightly different deal. "When their margins are bad I squeeze myself," says Mark. "But when they're fat I squeeze out an extra penny for Coast."

But as much as Mark adjusts, as much as he pulls the strings of forty-year-old relationships with refiners, the market continues to change. As the jam continues, people are looking to refiners for answers. The refineries sit at the boundary between consumers and producers—both extraordinarily powerful and surprisingly vulnerable—and at the center of the action. I ask Mark what he thinks about them. "How are prices set?" he says with exasperation. "There are eleven studies which show there isn't a conspiracy. It's hard for me to believe there would be a conspiracy. Chevron, Shell, Exxon—they hate each other. It's like war daily. For them to collude is insanity, but people believe what they want to believe."

Mark suspects it will be a long and bumpy ride as the system gets more and more burdened. "We've all seen a train wreck coming and we didn't do anything." He has an hour more work to do tonight, and it's already after seven. "I tell my wife that Coast is my mistress," he says cheerfully. He relaxes during his half hour commute home, he says, in a V-8 Lincoln "that can go 140." But, he says, his next car is going to be a Prius.

## 3 REFINERY *THIRTY SECONDS OF PANIC*

It's 8:00 A.M., and I'm standing in front of BP's refinery in Carson, south of Los Angeles, trying to figure out how to get inside. In this city of cars, I'd expect the door to the holy font of gasoline would be marked, if not by a faux gothic gate, at least by some Disney-style turrets. Instead there is a high unbroken wall, with silver towers and mounds of steam appearing above it. The wall goes on and on and the landscape behind it changes, now tanks, now towers, now more steam. When I do find a gate, it is the wrong one. And in the ten minutes it takes me to drive to the right entrance, the place pumps out another 45,000 gallons of gasoline.

If all goes well, by this time tomorrow morning the towers and steam will have turned 275,000 barrels of crude oil into 6.3 million gallons of gasoline, 1.9 million gallons of diesel, and 1.7 million gallons of jet fuel. And, according to the latest EPA numbers, the refinery will also have produced 1,523 pounds of toxic chemicals.

Inside the gate, I have to drive through a small maze to get to the main office, hunkering in the shadows between the pipes and the highway. And I'm still not in when I present myself to the receptionist. First, she says, I have to watch a video and take a quiz. "Safety glasses must meet ANSI 287.1 standards." My image of a refinery is something like those old cartoons of the boilers that huff and puff, straining at the seams, always ready to blow. The video addresses that issue: In the rare event of an emergency, three beeps will signify a "unit alarm," and one should leave the unit immediately. An all-clear will be a fifteen-second continuous blast. The quiz is mostly about the alarms. Once I've passed, I meet Walter Neil, head of community relations for the refinery.

When I called, Walter was open to the idea of me hanging around the refinery for two days. "Come see the best," he said jovially. In person he is angular and sincere. A former music major, he's worked in L.A. re-fineries for more than twenty years, first as an operator, then as a safety supervisor. Gradually he became the refinery's public face. "I knew what they were thinking of us," he says. "We were big, horrible, polluting, and stinky. Money-hungry ogres hiding behind a fence." Walter is on every possible civic and charitable organization in Carson.

This refinery was built far beyond the outskirts of the city in 1923, near the unlovely oil wells and seedy coast of Long Beach. But now the facility sits just off the 405 freeway, in the middle of the sprawling lifestyle created by its own gasoline. Refineries with their bulky, scab-bily painted equipment, their frightening flares and toxic possibilities, are antique vestiges of the industrial age, but we have not outgrown them, we've just grown around them. Being the face of a refinery is difficult because an oil company's reputation is only as good as its last disaster. In 2002, air quality inspectors arrived with a warrant and sheriff's deputies to audit plant emissions before filing lawsuits totaling more than $500 million, citing the plant for thousands of air quality violations between 1994 and 2002, including a release of hydrogen sulfide gas that caused elementary school students to be sealed in their classrooms. BP claimed it had trusted its inspections to an outside contractor that had failed to keep up. Without admitting guilt, the company agreed to pay an $81 million settlement. According to regu-lators, the plant is now "more cooperative and more conscientious," and has more staff working on environmental compliance than other refiners in the area.

Walter gets ready to take me on a van tour. "When you take the mys-tery out of a place like this, people are more comfortable."

Entering a refinery is leaving the familiar world of gas stations, tanker trucks, and highways behind. Here, in any direction, the view is occluded by a horizontal pipe—or fifty—about twenty feet off the ground. The refinery reminds me of the crystal gardens kids make with salt and ammonia, which at first appear chaotic but on close inspection turn out to have a rigid organizational structure. But I can't find that un-derlying logic in the pipes, the steam trails, and the beginnings of things and ends of other things that extend out of my sight. I can't even throw my neck back far enough to see the metal casks that are five stories high. I feel like an ant in a maze designed for rats. "It all looks like a bunch of spaghetti and meatballs," says Walter, "but everybody who works here knows what goes in and where it goes."

Refineries are molecular butchers, dissembling crude oil and shaping it into smaller, usable components. Crude arrives as a stew of hydrocarbon chains—some as short and gassy as methane, which consists of 1 carbon atom and 4 hydrogen atoms, and some as long and heavily sludgy as the asphaltenes, which can have 150 carbon atoms surrounded by messy scrums of hydrogen atoms. Mixed in you'll also find sulfur, salts, nitrogen, and metals. A refinery sorts these molecules by size and behavior and then cuts and re-forms as many as possible to make the 3- to 12-carbon molecular variety pack that is gasoline.[1] Sorting requires fractionating towers to separate the components by weight, and a whole variety of other vessels with catalysts, vacuums, re-formers, and compressors to re-shape the molecules. The key ingredient is steam—a million pounds an hour, one-and-a-half gallons of water for every gallon of crude.

**Crude oil** arrives on tankers in the port of Long Beach, entering the southern end of the refinery by pipeline. Walter and I start there and head north, past the fractionating towers that sort the hydrocarbon strings by their boiling point. There are dozens, maybe hundreds of towers in the refinery, all looking like headless silver rockets with a spiral staircases twining around them. Inside each tower forty trays sit at different levels. When the steamy hydrocarbons enter as gases, they condense in the trays with the lightest molecules on top and the heaviest on the bottom. The streams of sorted hydrocarbons are taken to the next treatment, and the next. We pass pipes and towers and more pipes until we close in on the fluid catalytic

---

[1] Gasoline comes in different assortments of molecules, from regular to high octane. High-octane gas tends to have more of the longer (more expensive) molecules. In an informal survey, three out of three refinery engineers put regular in their cars. Why? Because premium gasoline is about the engine, not the fuel or the driver. Higher-performance cars use more compression in their engines, and higher-octane gasoline, with its longer carbon chains, can withstand those pressures without "knocking." When fuel burns in an engine's cylinders, it ideally burns in a long, progressive flame out from the spark—like a fire moving across a dry field. But gasoline, particularly the lighter molecules, will also ignite under pressure. Knock happens when gasoline suddenly explodes from pressure, making a knocking noise and damaging the cylinder. So high-octane gas is only for high-performance cars; the rest of us can relax and buy regular.

cracker, the refinery's most important piece of equipment. The cat cracker uses steam, hydrogen, and a catalyst to break long hydrocarbon chains into short, neat gasoline molecules. It produces a quarter of the gasoline the city of Los Angeles uses daily, and it feeds a whole complex of towers and equipment that finish the gasoline. "I call them the children of the cat cracker," Walter says, "All of them serve that big monster."

I had hoped this landmark would reveal some of its alchemy in a visual inspection. (I had a fantasy that I would be able to somehow see the cracking and re-forming of molecules into rings and branches, as if clowns were twisting animals from balloons.) But the cat cracker, like everything else, is wrapped in steel and rivets and painted pale blue and dun. Disasters, Walter says, can be visible, odorous, and expensive. "If there's a power failure or an operator error and you reverse the flow, you get a big yellow cloud. Oily and horrible." We spin past the refinery's own fire department, its medical center, the generator. Workers in jumpsuits appear and disappear between the pipes.

At the far corner of the 630-acre plant sits the coker, cooking the sludgiest collection of leftover heavy molecules into chunks of black carbon, which are then sold as cheap fuel to China. Charcoal briquettes the size of small cars tumble down a slide to a conveyor belt, making a loud, irregular rhythm of glunks and wumps. The coker appears to date from the medieval era, but it may soon be reconfigured to become a post–space age energy source. BP plans to revamp this process with a $1 billion plant that will turn the coke into a fuel gas and use it to produce hydrogen and electricity, while storing greenhouse gases underground in nearby oil fields. Not only is BP acknowledging global warming—something U.S. oil companies have been loath to do—it appears to be positioning itself to profit from greenhouse gas regulation. The plant is a sort of paradigm-shifting move—in terms of both the chemistry of refineries and also in the politics and economics of global warming. The proposed plant is one reason I wanted to visit the Carson refinery.

It's 10:08 and we're driving toward the hydrocracker (another big cracker reactor that produces diesel) when the sirens blow. *Beep. Beep. Beep.* I remember that from the safety video! That's a unit alarm. Workers appear in streams among the equipment. They are wearing hard hats and carrying their tools and walking steadily and purposefully. The streams join up and then part in front of our van. Dozens and dozens of workers walk past us. They aren't rushing, but they aren't dallying either. "Don't worry," says Walter. I haven't gotten around to worrying yet, but I should be. Firetrucks arrive. Almost a year earlier, in March 2005, a catastrophic explosion and fire at BP's refinery in Texas City, Texas, killed 15 people and injured 170.

Walter keeps talking, his voice well modulated and calm, but his rapid eye movements suggest this situation is awkward. "I've never had this happen on a tour before," he says, fiddling with the shifter on the van. When the crowd eases, he backs up and we leave the area. Jets of steam are shooting from the side of the generator plant like an overheated teakettle. Above us, a flare wavers with a peachy flame. Walter turns another corner: another flare. And then three more. When all is well in the refinery, the flares are not lit. Walter thinks the generator went down, and the effects of that tipped from one process to the next, until the unprocessed gases in the system went to the flares, where they are being burned.

Flares are the defining symbol of the refinery: To the people who work there, the flare is a safety device, removing harmful or explosive chemicals from pipes by burning them before they escape to the air. To the people who live outside, the flares are proof of pollution, their witchy flames ravenous and scary. When the flares go off, everyone starts calling the refinery. The neighbors call to see what's going on; the regulators call; and the fuel traders call to see if there's going to be a price spike. Walter realizes he's forgotten his pager, and he's surely getting paged, so we go back to the office.

In front of the office we come upon the environmental management crew standing in a clot, squinting at the flares. Today is not a "Really Good Day." The oil industry loves quantification and incentives, and engineers live by spreadsheets, hence the enumeration, evaluation, and fetishization of "Really Good Days" at BP. When Carson meets its safety, environmental, and business targets for a twenty-four-hour period, it's had a, well, RGD. The Carson refinery needs to have 82 percent RGDs, and everybody's bonus depends on it. Whatever just happened—whether it's a big accident or a smaller one—the refinery is now out of equilibrium, the Good Day has turned.

**The shutdown** changed my plans to visit the refinery as the center of the oil industry, the midpoint between the consumer and the oil fields, the place where gasoline and gas prices come from, and a future way station for greenhouse gases. But the shutdown also offered me the chance to see how refineries really work. "I joke that working in a refinery is 2,793 hours of ass time in front of a computer and thirty minutes of sheer terror," says refinery workers' union rep, Dave Campbell. "And you don't get paid for the time in front of the computer. You're getting paid for tolerating those thirty minutes of terror." I had been presented with a ticket to look at those thirty minutes.

The first stop on the shutdown tour is the central control room. Hidden behind a nondescript door, the room is dark, lighted only by a rough hexagon of computer screens, forty-one in all, winking with colored schematics of every process in the refinery. In the dark, the people who control the computers are negative silhouettes against the bright screens. "I say it's like a video game with real results," says Gina, a tall, athletic woman with what I take to be an Oklahoma accent who monitors the cat cracker. I can't see her face in the dark, but she starts spouting homespun wisdom adapted to hydrocarbons in the midst of this NASA-like array. "I make a lot of horse analogies," says Gina. "This place is like a shy horse. Just when you think you've got 'em used to sugar cubes, a rattlesnake will bite 'em and then they kick you in the face."

At around 10:00 A.M., she says, one of the refinery's generators went down while another was being repaired, leaving the unit without backup electricity. A three-second power dip hit the substations to the hydrocracker, and its pumps and compressors lost power and shut down—that was when the unit alarm went off and all the workers started filing out. That, in turn, shut down the hydrogen plant, which produces hydrogen and steam, and the butane and pentane units. Refineries run on steam. Without it, a refinery stops, chemicals stop flowing, and things start to go wrong. The pressure began dropping across the whole system.

Around the control room, other unit technicians were furiously protecting their units. "When the alarms are going off, there's a stress level," she says. "You gotta be careful." Gina and her coworker Chris could see that the shutdowns were moving toward the catalytic cracker. "We had to get the steam refocused to the blowers," says Gina, "so we started shutting this whole section down." She points at some blips on the screen. They watched the temperature rise in the section of the unit that needs cool air. The temperature is supposed to be around 500 degrees F, but it was rising toward 640 degrees. If the process got too hot it would stop flowing, which would mean the whole unit would have to shut down. The cat cracker would be gone for at least a day, and with it, a quarter of Los Angeles's gasoline supply. They continued shutting things down. "We just kept shutting everything down, trying to get all the steam we could to that blower." In the end, disaster was averted. "Actually I say you don't run this place," says Gina. "It's like a herd of wild horses and you try to guide 'em. And you got 'em going basically where you want and then one runs astray and you got to get it back."

I'd like to see the cat cracker, the "monster" that keeps a quarter of L.A. driving. Jennifer Mederos is a young process engineer on the unit who agrees to take me. In her blue jumpsuit, Jennifer is earnest, and her hair is pulled back tightly from her face. On the wall near her desk there's a Xeroxed cartoon showing cats, as in kitty cats, going into the cat cracker and coming out as "cat ends," or little cartoon cat butts, but I never see Jennifer in the giddy state the drawing suggests. Her work requires very close attention to the details and quirks of the monster, requiring large spreadsheets and boggling schematics of its parts.

"Fun it wasn't," she says of the shutdown, gathering papers on her desk. "Everybody's unit goes crazy. Everybody's under stress and the place scrambles to get back." We leave the oatmeal-colored bunker where the engineers are and walk among the pipes, which are dripping water and huffing steam. The ground is swept as clean as a park in Japan, without traces of oil. The high-pitched screech of the pipes and blowers shears through my earplugs.

The cat cracker was originally built in the 1950s, and to visit it, we ride in a very old green elevator ornamented with inscrutable vintage graffiti of a pig with an upraised fist. On the third story, we step out onto a platform about the size of a basketball court, but crowded with truck-size casks and smelling of boiling potatoes. Everything is held together with bolts the size of Dixie cups. Jennifer points out the parts of the cracker, but it's difficult to hear her amid the roar of the unit. If I could stand back from the platform at the proper height and distance, the metal tubes around me would look like a large chemistry set covered in multiple layers of scaly paint.

Gasoil, a refinery product that is heavier than diesel, comes into the cracker from the side in pipes the size of furnace ducts. The gasoil mixes with a metal catalyst resembling a very fine sand that has been heated to

1,269 degrees F, so that it flows like a fluid. The gasoil and catalyst spend a few seconds rising in a tube while a chemical reaction converts about half of the gasoil to gasoline molecules. At this temperature the molecules are in a gas form, so they are sucked out of the top of the tube by a cyclone and taken to a fractionating tower, where the heavy gasoline and the light gasoline are separated before being taken away to be further modified into finished gasoline.

Back in the cracker, the catalyst is now covered in carbon, and it is blown out into the center of a beehive-shape cask with air so that the carbon vaporizes to become carbon dioxide, water, and pollutants like nitrous oxide and hydrogen sulfide, leaving the catalyst "regenerated" and ready for more action. Eighty tons of catalyst runs through the beast every minute. If the shutdown had hit the cat cracker, the catalyst would have stopped moving, which would have taken more than a day to clear up. In the worst-case scenario, the motion of the catalyst could have reversed, causing the big horrible yellow cloud that Walter talked about.

The cat cracker proceeds as rationally as the formula for a chemical reaction, but it also has a cranky, devious side. Jennifer spends a fair amount of time sleuthing out the cause of anomalies. Why are there small amounts of hydrocarbons in one of the electrostatic precipitators that remove particles and smog-forming chemicals? She spends a day determining whether it's an emergency (it's not) and then a week doing detective work to figure out where it came from. She runs different chemicals through the cracker to see if they affect it, and she pulls out old maps of the unit, searching for its secrets. "I like climbing around the unit," she says. "I go to the junctions and see the pressures and the flow rates."

Later, at the unit's morning meeting, Jennifer and the unit's superintendent, also a young woman, meet with the plant's operators. The operators are mostly union guys in their forties and fifties, and they treat the cat cracker as a somewhat lovable old fussbudget. They say the unit "likes" things and dislikes others. They spend a long time talking about variables—heat in one place, speed in another, temperatures rising and falling, adding chemicals to accelerate the process, and variations in the feed. While refineries appear unchanging, the processes inside are in a constant state of flux and occasionally entirely off kilter. "It wasn't bad as upsets go," says one operator. "We got control within an hour."

In the end, the worst part of the emergency was that one of the operators was badly hurt. When a pump needed to be restarted by hand, he jumped onto a bicycle with a wrench to rush to another part of the unit. The wrench fell out of his basket and between the spokes of his bike, flipping him over the handlebars and sending him to the emergency room with injuries to his face.

**At the end** of the day the flares are still burning, and they're Sue Sharp's problem. Sue is Carson's environmental manager. In her thirties, she wears an all-American red gingham shirt and a blond bob, but on the wall behind her hangs a map of the places she's worked for BP, including a stint in the London headquarters, Alaska's North Slope, Azerbaijan, Venezuela, Egypt, and South Africa. As I talk with her, two decorations in her office seem to frame her head: a trophy of a pipe valve celebrating the mapping of 357,000 valves and joints, and a copy of Dr. Seuss's environmentalist allegory *The Lorax*. Sue was brought to L.A. to clean up, both literally and figuratively, after the regulators raided the plant.

"Today is not a Great Day," she says. She was sitting at her computer when word of the shutdown came on the radio. She instantly went into action because a new regulation, as of January 2006, requires that refineries limit flares and report them. She immediately called in the flares to the local air quality regulator (the Air Quality Management District, or AQMD), the Office of Emergency Response, and a national response center. Because the flares were going for more than half an hour, she dispatched her team to get samples of the gases in the pipe to send away for analysis. In the meantime she sat at her desk, looking at the flow rate of the flares on her computer screen. "There were pages coming in and there were pages about to come in." Members of the community were calling both BP and the regulator. "And they have to tell people the refinery isn't blowing up." She took a drive through the plant to see what she could smell: gasoline, tar, asphalt, and mercaptans—the odor added to natural gas. And as soon as the emergency was over, she had to begin an investigation and prepare for a follow-up visit from an AQMD inspector the next day.

Sue says she was attracted to environmental engineering because she likes its "gray areas," and in L.A., with its combination of smog and tough politics, she has found a big gray area. Sue describes the regulations here as "the most stringent in the world."

The amount of pollution a refinery emits depends partly on where it is, partly on who's regulating it, and partly on how hard they're trying. Los Angeles has become a leader in emissions regulation, a place where activists push the boundaries to make changes across the United States. "It's a trickle-up approach," says Julia May, who spent seventeen years as a staff scientist at the environmental advocacy group Communities for a Better Environment. "We get regulations adopted on refineries here with the local agencies. There's always a huge fight. But once you show it can be done these regulations spread across the country."

Making and enforcing rules at refineries is contentious, and it seems that everyone is at everyone else's throat. Both the refiners' association and community and environmental groups have sued L.A.'s AQMD either to prevent or to compel the enforcement of air regulations. In return, the AQMD has sued the refiners. And in 1997 the U.S. Environmental Protection Agency (EPA) did an audit criticizing the AQMD for timid enforcement, poor measurement, and low fines. Meanwhile, the oil companies do not all agree with one another, and the unions are asked to take sides with the refiners or with the communities. It gets very ugly and very confusing.

The flare rules that Sue needs to follow today are the result of that contentious process, and they show what a smoggy science regulating refineries is. Rather than measuring the chemicals that escape from a refinery with electronic equipment or sniffers, regulators catalog the refinery's equipment, pipes, valves, and tanks and multiply them by standardized estimates, called emissions factors. Emissions factors can be very inaccurate.

In early 2006 the EPA realized that its emissions factors for refinery flares represented about one-fiftieth of the actual potential emissions from them. The miscalculation caused at least 765,000 tons of uncontrolled emissions in Texas, Pennsylvania, and California.

And emissions factors leave a lot of room for interpretation. In California, for example, regulators assume that a flare will destroy 99 percent of the volatile organic chemicals (VOCs) coming out of a stack, but Texas assumes the same flare removes only 98 percent of the VOCs. Move a refinery from California to Texas, then, and its flare emissions estimate would double from 1 percent to 2 percent. "Right away emissions in California appear to be half those in Texas," says Eric Schaeffer, a former EPA regulator who's now with the Center for Environmental Integrity in Washington, DC. "It's a compounding problem, and there are huge gaps between the estimates and what actually comes off the stacks."

Flares are considered safety equipment, which means refineries would, in theory, use them only during emergencies. But when L.A. regulators studied the flares, they found that between 1999 and 2003, 83 percent of the gases vented to the flares were nonemergency. The refiners' association fought the flaring rule with gusto, but during the course of the study, the refineries themselves reduced flaring by about half. And curiously, some refiners found that installing equipment to capture the gases that would have gone to the flares actually made their operations more profitable. One California refinery made back its investment in capturing flare gases within two years and then actually saved more than $50 million a year. In the nitty-gritty details, refinery regulation seems to be a political art rather than an objective science.

Sue says that preventing emissions is part of the refinery's "social license to operate." She'll file a report on the morning's flares, adding it to the three hundred other reports and audits she'll do this year. She doesn't imagine that her workload will get lighter or environmental regulations will ease. She's making budgets for 2012 with the assumption that standards and scrutiny will only get tougher. Between now and 2009, the refinery has $300 million in projects aimed at environmental compliance and increased efficiency, not counting the $1 billion plan to store the carbon dioxide. Regulations on greenhouse gases are inevitable, she says. "My question is: What's the price?"

**That afternoon** I get in my rental car and turn the key. Even though it's cold and damp, the engine starts with a thrum of perfectly timed explosions. I nudge out onto the crowded freeway, come to a stop, and have time to stare at the refinery's flares and big doughnuts of steam and wonder what I'm breathing.

The health effects of refineries are hard to quantify. A project on the Gulf Coast of Texas has attempted to track refinery emissions and community health. "We've found cancer clusters," says Dr. Winnie Hamilton, director of the environmental health program at Baylor College of Medicine, "but the geographic resolution isn't high enough. Sometimes the lung cancer corresponds to a higher smoking rate. We do a little sampling and we find higher levels of benzene in working-class communities. But that's partly because they are more likely to use mothballs."

One big problem about refineries, says Hamilton, is that people who live nearby feel they have no control over them. Disasters, which can be horrible, are a case in point. "During a shelter in place [when refinery alarms go off and residents are told to seal their houses], people don't know if they're going to die, and sometimes they're separated from their children and the stress is immense." The question for the future, she says, is trying to determine if the combined stresses of toxic exposure, poverty, and feeling a lack of control affect people's health in ways that compound each other.

Living next to a refinery is being neighbors with the unknown. "We want to know what's going on over the fence and nobody's been able to answer that with a cumulative effects study," says Roye Love, a citizen of Carson who's on the city's environmental commission. "I think our community is overly impacted, and I'm saying why should we pay the price?" Refineries, like toxic waste dumps, landfills, and nuclear power plants, bring down property values, but they also bring in tax revenue.

Sitting in the traffic jam, I was inclined to focus my fears on the refinery, but meanwhile the engine of my rental Chevy is furiously breaking apart the neat carbon chains of gasoline and recombining them to form all sorts of pollutants. To go a single mile, this car will use 6 ounces of gasoline, and it'll puff out of the tailpipe 2.8 grams of hydrocarbons, 1 gram of nitrogen oxides ($NO_x$), nearly 1 pound of carbon dioxide, and 21 grams of toxic carbon monoxide.[2]

---

[2] According to the EPA, an average driver burning 581 gallons of gasoline a year will produce 77 pounds of hydrocarbons, 575 pounds of carbon monoxide, 38.2 pounds of nitrogen oxides, and 11,450 pounds of carbon dioxide ($CO_2$).

A gallon of gasoline produces 19.5 pounds of $CO_2$—imagine a large Thanksgiving turkey squeezing through your tailpipe. Gasoline itself only weighs 6 pounds to the gallon, so the $CO_2$ figures have always struck me as ungainly. But once the carbon atoms are liberated from their hydrocarbon chains, they team up with double oxygen atoms, and that's where the extra pounds come in.

On a normal day, the pollution produced by refineries pales next to what comes out of cars and trucks. In some parts of the L.A. Basin, the "estimated lifetime cancer risks" from air pollution are one thousand times the amount the federal Clean Air Act permits. And while there is a lot of industry in the area, about 70 percent of the excess risk comes from cars and trucks. The people who absorb these risks—the people who really pay—tend to be poor and disadvantaged. And for children, living within 250 feet of a major roadway raises their chance of getting asthma by about 50 percent, according to a recent study of 5,000 kids in L.A.

Gasoline sold in L.A. has been reformulated to reduce pollution; the reformulation is costing me an extra 3 to 10 cents per gallon. Still, I'm not paying for the pollution I emit. If I were a refinery, I'd have to buy credits to emit $NO_x$ and $SO_x$ (sulfur oxide). That single gram of $NO_x$, for example, would run me 4 cents at 2000 prices (when emissions credits happened to be very expensive), meaning that for that pollutant alone, I could be paying an extra $1 a gallon.

For as long as we've been using gasoline in the United States, we've defined it by its price at the pump, not its costs to society. Only recently, the cost of our overextended infrastructure has begun to creep into the price at the pump. Later that night, I drive past the Carson refinery in the rain. All of the flares are still going—spookily wavering amid the lights of the fractionating towers and the rainy fog. I wonder if prices in L.A. rose as a result of the shutdown. In 2005 big refinery outages here caused spikes of between 10 and 15 cents a gallon.

A few years ago refinery outages didn't mean much—there was too much gasoline on the market. Oil companies overbuilt refineries in the 1960s and 1970s, expecting that Americans would continue using more and more gasoline. By 1981 the United States had 324 refineries. But after the oil crisis of the 1970s, conservation and smaller cars made the demand for gasoline drop, which meant refiners struggled to make a profit. During the 1990s many smaller refineries closed and large oil companies merged. Now there are only 149 refineries in the United States, and they are concentrated in fewer hands, while demand for gas is so high that the country has to import gasoline that's been refined abroad.

There are two opposing theories of gasoline in this country. One says that the market is wise and high prices are a signal that everyone obeys. In 2006, as gas prices headed toward $3 per gallon, the President's Council of Economic Advisors echoed this thought, saying that increased scarcity of gasoline and rising prices would, over time, lead households to conserve. The other theory of gasoline is that lower prices are better prices, because consumers can't pay high prices and they can't be expected to use less gas. This is typified by President Bush's request, in April 2006, as gas prices passed $3, that states relax air quality standards, a move that might reduce prices (and also encourage more driving) but would shift the cost to the coughing kids by the freeway. A group of senators advocated lifting the gas tax, which would also have the undesirable effect of stimulating demand.

It was the same old talk that has been going on since the big energy crisis in the 1970s. While some politicians called for investigations of rigged pricing, others were working to relax refinery environmental regulations, which they said would mean that U.S. refiners would build more refineries and increase the supply of gasoline, lowering prices. Studies by the EPA, though, suggested that it wasn't environmental considerations that were limiting the building of new refineries, as much as economic ones. In the last twenty-five years, only one refiner has requested a permit to build a new facility. For one thing, refineries are usually positioned on land near waterways, which is now valuable to developers. For another, new refineries cost billions of dollars to build.

As crude oil prices rose, oil companies were reporting enormous profits. Most were coming from their "upstream" divisions, which produce crude oil and sell it at market prices, rather than their "downstream" operations, which do the refining and selling of gasoline. In 2005, BP's upstream profits were six times their downstream ones. Between 2003 and 2006, refinery margins rose dramatically, but those increases are calculated in pennies per gallon, while the bulk of the cost of gasoline is still the crude oil. On the day I'm in L.A., prices are $2.48 a gallon—$1.43 of that is related to the cost of buying and transporting crude. Another 54 cents go to taxes, and another 54 cents are refinery cost and profit, according to the California Energy Commission.

Industry critics said refiners were deliberately limiting their production capacity and stifling competition to keep prices high. The relationship between supply and demand was so tight that some energy economists issued a warning. "If consumers' demand is not sensitive to price *and* refiners are producing near their capacity constraints then it is likely to be profitable for a refiner to reduce production to drive up prices, possibly even if that refiner doesn't have a large market share," said a report cowritten by UC Berkeley's Severin Borenstein. When consumers can't use less gas, and refiners don't have to compete for business, we have something that is not "gouging," nor a cartel, nor a competitive market. Who will fix it?

**The next** morning I am back at the refinery to learn how it makes money. Yesterday's shutdown will be among the year's top five incidents, says Ken Cole, the plant's optimization manager. Ken is a compact man who shaves his head; he seems to appreciate the jumpsuit's utilitarian charm. Ken's job is coaxing the refinery to make more gasoline while shoring up safety and environmental issues. Debottlenecking, he calls it. Ken sometimes refers to the refinery as "the beast," as in "There's a tool for optimizing the beast." Ken crawls around the refinery through spreadsheets and computer programs.

As we walk in the shadows of the pipes toward his office, Ken says the refining business is going through a "step change"—a huge upset, a break with the past, and a rewriting of the rules. Ken's first engineering job in a refinery was in Coryton, England, in 1989. Immediately, there were layoffs because global demand for fuel was so low and refining profits were feeble. "The name of the game was lower costs. You had to be profitable when your absolute margin available was $1 per barrel." (That translates to being able to make a gallon of gas for less than 3 cents.) Many refineries in Britain and the United States closed, oil companies merged, and nobody invested money in their plants. Until the late 1990s BP viewed its refineries as suppliers for its stations, not as profit makers.

That changed after 2003, when China and India started buying more oil. "Global demand for oil has grown and almost outstripped supply. Any short-term effect like shutdowns—like today—this could affect the market. Prices are volatile, says Ken. He adds, "Obviously, in the last two years, profit margins are unprecedented." BP's first-quarter trading update for 2006 shows that the company's profit margin per barrel for its refineries worldwide was $6.38, though on the West Coast of the United States, it was $11.22, after hitting a high of $17.57 for a quarter in 2005. Globally, refineries were using 97 percent of their capacity, so supply wasn't just tight here; it was tight all over the world.

And when the company's strategists look at the future, they see more of more of the same. "Even if the economy turns down, and people get fuel-efficient cars, we couldn't find a scenario where demand will fall here on the West Coast, just because of population growth. Growth in demand is outstripping even our most aggressive strategies." By 2014 demand is expected to grow by another 25 percent.

For Ken, high demand makes work tougher. After two decades of layoffs, trained engineers are scarce, and he is spending some of his time on a big recruitment drive.

When a refinery is running near the peak of its capacity, it must walk a thin line between producing more and sacrificing profits. Gasoline is blended into a mix of molecules that meets air quality standards and has the correct amount of octane. Ideally, the blenders can choose among the streams of gasoline in the plant to make the cheapest possible blend of gas. But when the plant is going all-out, the processors start to fill up, which means they have to use expensive ingredients in cheap gas. "There comes a point when everything fills up," Ken says, "and then you're getting about 85 percent [of the potential income from a barrel of crude]."

So Ken and his team apply themselves to eliminating those bottlenecks where the processors fill up. They move parts closer together, change catalysts and mixing patterns. One percent at a time they remove the constraints, and when they're done they go back and start the whole process over again. They install more equipment, faster fans, more efficient heaters. This concept, called refinery creep, explains how refineries grow. Between 1990 and 1998 the production of the average refinery in the United States crept up by 28 percent.

Another way to make profits is to use cheaper, lower-quality crude oils. Good light oil has become scarce and expensive, but sulfurous heavy oil is plentiful and selling at a discount of as much as $15 a barrel less than premium. Changing from good oil to cheaper oil cannot be done quickly or thoughtlessly—you need equipment to remove sulfur and salts. "Over the last few years we've been pushing the boundaries on Arab crudes. The salt makes thin metals on the unit, and you have to go in and fix them." These cheaper, high-sulfur crudes will be even more plentiful in the future, so being able to process them is a competitive advantage. This is a challenge. "You have to understand the boundaries and manage the tension," says Ken. "Without action you pay big."

To prepare for a future of heavy crudes, the refinery is replacing a fifty-year-old vacuum tower. The old tower broke apart heavy oil, but always ended up with a large quantity of sludgy leftovers that had to be sold as cheap fuel. The new tower will use a vacuum filled with hot steam to crack the long sludgy molecules into shorter chains of gasoil, which can then be refined into gasoline or diesel. Ken shows me the site, where workers are crawling around on a five-story scaffold, welding large rings of steel into a tall chamber. Like a Lego toy, the new tower will be plugged into the refinery's pipe grid to replace the old one. The new tower will create an extra 6,000 barrels per day of gasoil from the same amount of crude, bringing the refinery profits of $50 million a year.

Wringing more from every barrel of oil is a national trend. The amount of oil products coming out of American refineries now exceeds the amount of crude oil going in by more than a million barrels a day, a phenomenon called processing gain. Processing gain shows how innovation can reduce the need for more crude oil. (Drilling in the Alaska National Wildlife Refuge, for example, would yield almost the same amount of crude—between 0.6 and 1.6 million barrels a day, according to estimates by the U.S. Geological Survey.) Refiners have to make big investments—like the vacuum tower—to increase processing gain.

Another profitable trick is turning cheap products into expensive ones. Carson sells its coke—those giant charcoal briquets—to China cheaply. Managers wanted to find a way to make it into something more expensive, and they wanted to take advantage of California's plan to limit greenhouse gases and let producers trade credits. And so they decided to build the new electrical generation plant. By pumping the greenhouse gases into nearby old oil reservoirs, not only would they store them out of sight, but they also might increase the flow of oil coming out of the wells. And when trading in greenhouse gas credits starts, the company will be positioned to make more money. "This is more expensive, but the economics of trading gases could make it worthwhile," says Ken.

BP's project is an interesting model of how environmental regulations might be good for business. The politics of greenhouse gases in the United States made similar government projects slow, expensive, and tinged by futility. BP's power plant is bigger, faster, and privately funded.

"Why should the taxpayer pay a billion dollars for FutureGen [an expensive federal clean coal project] when BP will spend that on this project?" says Dave Simbeck, a consultant with SFA Pacific, who believes that the BP project could potentially turn a profit and be a model for disposing of greenhouse gases. "There seems to be a disconnect. It is much better to create incentives so business finds the solution."

I asked Ken what he was going to do when greenhouse gases were limited. Was it going to be a big burden? "$CO_2$ is a spreadsheet item." He shrugged. "It becomes just another economic factor. It's just another variable in your decisions, and you can put a number on it and financialize it."

For me, this is the lesson of the refinery: With fuel, you always pay—whether it's higher prices at the pump or more pollution or more uncertainty. The big question is: *Who* pays?

As we sit in front of Ken's computer, I imagine a big machine of interdependent gears and economies and chemical reactions reaching out toward the gas pedals of those cars on the freeway, and back to oil fields far away. I find this vision strangely comforting, because it suggests that consumers could make rational choices to prepare—optimize the beast—for five, ten, and twenty years in the future. But will we? Ken talks about where things are going. "It'll need a step change to really change behaviors. That's the dilemma the world's got, right? Do we just continue as we go until it becomes unfeasible, or do we do something?"

Just before I leave the refinery, Ken opens a spreadsheet and reads the names of yesterday's crude oils: ANS, Hungo, Basrah, Escalante, Oman. At first I don't recognize their names, but later I realize they are specific crude oil fields, and oil, like wine, is named by its appellation. ANS is from Alaska's North Slope and Hungo is a field off the coast of Angola. Basrah is in Iraq, Escalante in Argentina. Oman is in the Persian Gulf. A whole world of crude is waiting.

**4** **DRILLING RIG** *LIVING BY THE DRILL BIT IN TEXAS*

After dark, East Texas radiates heat. Out the car window: singing insects and silhouettes of scrub brush. I turn off the highway at a town named Dew and head toward one called Luna, scanning the cow pastures for the first drilling rig.

A fuzz of halogen light appears with a tower in the middle. I brake to stare. A heavy industrial exoskeleton of crossbeams and wires, the 150-foot-tall derrick has trailers gathered around its base. The business end of the rig—that chomping drill bit—is buried under the soil and out of sight. The angled legs of the derrick have an insectlike quality, as if a monster mosquito were crouching over the skin of a giant, hindquarters in the air, head buried in a vein.

According to the weekly rig count, 27 of the nation's 980 gas rigs are at work in Freestone County, where there is a boom in natural gas drilling. They should not be hard to miss. "You could chuck a rock between 'em," said a local to assure me that what we had here was a genuine boom situation. "They gotta stick a lotta straws in the soda pop out there."

A boom is a relative thing. Between 1901 and the 1950s, oil boom-towns sprang out of Texas's geology and culture, moving on as soon as the oil was exhausted. A hundred miles northeast of where I'm driving, the boom to end all booms began on December 28, 1930, when a gusher of hot oil shot out of a hole in the fields of Lou Della Crim's farm in Kilgore. The oil came from the "Black Giant," the largest reservoir ever found in the United States—forty-two miles long and shaped like a side of beef. Before a year was out there were 3,607 wells drilled in the field, and the derricks were so thick you could travel for miles from derrick to derrick without ever touching the ground.

The gushers, the derricks, and the oil are all gone from East Texas now. What's left under this scrub is natural gas, which is formed from oil, and is found and drilled in the same manner. I wanted to spend time on an oil rig, but 90 percent of all U.S. drilling rigs are looking for natural gas now, not oil, so the odds were on gas. In this atmosphere, a gathering of a few dozen rigs somewhere around Dew and Luna can reasonably be called a boom, and I've come here to learn what makes a boom—both underground and above.

For the first fifty-something years of the twentieth century, finding oil in Texas was essentially luck: a sophisticated variation of those game shows where contestants choose the prize behind door number one, two, or three. Most people chose to drill in the wrong hill: 90 percent of wells were "dry holes"—empty.

Technology and economics long ago changed the odds and the game so much that Houstonians call the new oil and gas producers "mildcatters" instead of "wildcatters." The riddle of modern Texas is that the myth of the lucky wildcatter lives on, even when the men who lived that life are mostly dead.

A second derrick appears: my cue to turn right onto the access road to a third, where I hope to find a man named C. D. Roper, who's more or less dared me to come visit him.

In phone conversations, C. D. has told me he is a fourth-generation oilman tracing back to a great-granddad who was a blacksmith and part Cherokee Indian. "I had drill cuttings in my first baby rattle," he said the first time we talked. "My uncle made me a rattle out of a gourd and my grandpa decided it didn't make enough noise so he went out and got some Austin Chalk cuttings and put them in there." Austin Chalk is the name of a notorious limestone rock layer in East Texas. A well in Austin Chalk will appear to produce lots of oil, for a day or a week, but quickly peters out. "A sucker's play," C. D. called it. "My grandpa figured if I got close to Austin Chalk with that rattle, then I wouldn't need to touch the stuff later on. People have lost their ass in Chalk." Did he lose his ass in chalk anyway? I asked. "Yes. I did. I lost my shirt and I wasn't even spending." C. D. had a deep Texan drawl that sometimes crumbled, like something that had been dipped in batter and fried too long. "I've lost $200,000 so many times it seems reasonable," he said.

People who love oil, particularly Texans with romantic feelings for the stuff, often say they enjoy the smell of petroleum. C. D. belongs in this category. "I was like a well-trained ranch dog," he said one night. "Whenever my dad and grandpa were going out to visit a rig I'd sneak in the car. I loved the sound of the drilling rig. The smells." From that elemental obsession, he's spent his life in every facet of the oil business: from directional drilling, to geological analysis, to negotiating property rights, to writing software for the industry. "I love the process of getting it out of the ground. People think there's big caverns of oil and you just stick a straw in them, but it's really more like a sponge."

At the end of the road stands the derrick. I enter the square garrison of trailers around its base, park, and open my car door to a stunning wall of 100-degree heat and throbbing noise—*cachunk cachunk*. By the time I stand up this sound has segued into pounding vibrations beyond my hearing. I feel seasick.

The derrick has another white trailer lodged partway up it, as if a child jammed a toy in the rigging before heading off to supper. On the ground below it is bright as day, if daylight were green, and the shadows have thin green rinds. When I look down my feet appear foggy, as if I am watching them on a staticky TV screen. Leaning over to get a better look, I discover a flat mass of brown insects hopping in unison three inches off the ground. This is all a little more Alice in Wonderland than I had expected.

An old luxury car sits next to me, not a Cadillac but a Chrysler Fifth Avenue, with its hood up and a hatchet resting on the engine block. Who uses a hatchet to fix a car? And another thought, more alarming: Will I need this hatchet later? Perhaps this is a video game where the object is to collect weapons and tools as you go along. Then I high-step through the bugs toward the trailer labeled "Mud Logging."

The door of the trailer opens and a small, grayish man looks at me with hostility. I yell that I am looking for C. D. Roper. "He's not here," says the man, setting his mouth in a line that turns down at the ends. He wears wire bifocals that magnify his eyes so he looks like a belligerent bulldog. Frozen air curls out of the trailer. The bulldog turns away and gestures me inside disinterestedly. When the door wuffs shut, the trailer is immediately cold and almost quiet. In place of grasshoppers there is a highly polished, even lustrous, linoleum floor. With his back still turned, the man motions to a booth with screwed-down benches, saying "Pull up a seat." I choose to interpret this as irony, or perhaps a friendly joke. I wonder if I should say something positive about the extractive industries.

Texans can be defensive about oil because they assume the rest of the country does not approve. A prime example is R. L. Gaston, the amateur Texas historian who introduced me to C. D. Roper. Gaston was the author of a detailed two-part analysis of the relationship between Texas oil fields and variations in chicken fried steak. One afternoon, while attempting to read *The Elements of Petroleum Geology,* I procrastinated by hunting down Gaston's phone number and leaving a message on his answering machine. Gaston did not return my call for a few days, explaining later that I was from "Caliprunia" where they "are prejudiced against the extractive industries." Eventually he returned my call and we talked for an hour or so about Texas history. He started to talk about how much he loved oil. "They say oil rigs are a blight!" he said, apparently astounded that anything he'd admired since his earliest childhood, the fuel of cars and wellspring of manhood itself, as patriotic as chicken fried steak, could be seen negatively. "Ugly!" he exclaimed, as if the oil well's beauty was something that any chivalrous Texan would defend with his life. He suggested I call a man named C. D. Roper, who really loved oil. Now C. D. has slipped my Caliprunian grasp.

The bulldog harrumphs at his computer. I make small talk, mentally cataloging the contents of the trailer in case this is the closest I get to the drilling rig. Formica counters running on three sides of the trailer hold a microscope, a sink, two heat lamps, a pile of small trays, sieves, a funnel, a UV light with a viewing apparatus attached, an old PC, a cabinet filled with tubes and small electronics, stacks of paper with jagged lines on them, boxes filled with rows of tiny manila envelopes, a disassembled carburetor, a coffeemaker, and a large bottle of floor wax. At the end wall there are two bunk beds, one holding a guitar.

The bulldog's back is toward me. "Where's C. D.?" I ask, thinking I better get it over with if he has stood me up. "I am him," the bulldog says, folding himself double, slapping his knee and cackling. Then I recognize his voice, with all those crackly bits falling off the sides. His face crinkles, and he hops around the trailer slapping himself with glee. Of course, the hatchet and the car are his. So's the floor wax. He is delighted by my shock. "You look like a possum been eatin' a persimmon."

So ended C. D.'s final moments of silence. Over the next six days, he will talk about ice hockey, motorcycles, the two kids he raised by himself, the Grateful Dead, literature, Liberia, computer programming languages, petroleum geology, and cooking. He will describe at least five women as "She could go bear hunting with a switch." How he managed to sit in silence so long, pressure building, or why, will remain a mystery.

C. D. hands me a hard hat and tells me to go turn my car around. On a rig, all the cars point toward the road in case there's a blowout and everyone has to leave in a hurry. Blowouts: If the drill bit hits a pocket of high-pressure gas, it could overwhelm the safety equipment and "spurt out like a watermelon seed squeezed between your fingers." If a blowout catches fire, it becomes a tower of flame, possibly accompanied by poisonous hydrogen sulfide gas. Sometimes the pressurized gas picks up the derrick and hurls it somewhere. Same for the two miles of heavy pipe inside the well, which flies into the air and comes back down "like spaghetti." If a blowout occurs, there is no point in calling 911. Instead, call 1-800-BLOWOUT to summon a company called Boots and Coots. Blowouts are very uncommon but not unthinkable, so cars are ready for a quick getaway. So goes optimism on the rig—always forward into uncertainty, with flimsy, almost ritualistic hedges against disaster.

## Rig: Night One

When I come back in the trailer, the drill bit is precisely 10,028 feet below us, crunching downward at a rate of twenty-three feet per hour. C. D.'s official job on this rig is "mud logger," which means he creates a record of the progress of the bit through the rock layers below, samples the rock, and predicts where the bit is going. If you want to see what East Texas looks like to a drill bit, sit in C. D.'s trailer and watch him work.

In front of him is a chart showing the rock layers in this part of Texas. The names are reminiscent of suburban cul-de-sacs: Pecan Gap Chalk, Blossom Sand, Glen Rose, Rodessa, James Lime, Travis Peak. C. D. treats these layers as a sort of neighborhood, with a lot of familiarity and affection. "I think there's a pool of primordial knowledge, and some people stick a straw in it," he says later that night as he does a calculation. "I'm into the numbers. Numbers are the way to understand the oil fields." Nothing makes him happier than calculating accurately where one layer leaves off and another begins. C. D., though, is never just happy, he's usually "happier than a three-nutted tomcat."

East Texas, he explains, was a sea after the breakup of Pangaea 200 million years ago. As Pangaea split, North and South America separated, forming a huge ocean basin between them. The East Texas Basin was a subbasin of the larger Gulf Coast Basin. Over time, rivers carried silt and crushed rock into the basin, leaving behind layers—shale, sandstone, limestone, shale—like a giant sedimentary lasagna across the whole area. Some layers are famous for the fortunes they've made, such as the Woodbine Sand, which held the oil for the Black Giant and was only 3,500 feet deep. The bit on this rig is headed for the Bossier Sands, which are around 12,000 feet deep.

C. D. is buzzing. No sooner has he adjusted the electronics on the gas chromatograph than he's off to meddle with the carburetor for his Fifth Avenue. That gets abandoned when he calls his son. "Son. About the cat. Could you wash around his food bowl with Clorox? Cats track in germs." Floors in general are a recurring concern: When he's between tasks, C. D. sweeps the trailer. He does a calculation, muttering "Ninety-two, 64, I may be talking to myself; 15, 87, 6 . . . Why is this happening?" He takes a detour to explain why hydrogen sulfide gas is so dangerous. (When the gas is present in dangerous concentrations, your nose loses the ability to sense it. Then you die. At the first whiff of rotten eggs, scram.)

How did he get to this mud-logging hut? Growing up in Odessa, he was in thrall to oil and at the same time gifted in math. "They set my mind on fire for technical shit, and I haven't gotten over it since," he says. After his father died he ended up in reform school. He describes that period of his life as "feral." He was also in the marines, though he doesn't talk much about that time. In college, he studied nuclear physics and played guitar. He never became either geek or hippie, maintaining a certain solid Texan slightly miffed intensity. A career with an oil services company took him around the world, but he still returned to Texas to be a directional driller. When his wife left him with two toddlers, he adapted, hiring a nanny and getting a radiophone so he could raise the kids from the road. Sometimes he wrote software for the oil industry.

In between stories and work, he will suddenly drop everything and run outside to pee: a symptom of kidney problems. He returns, cracks a joke, and picks up the conversation precisely where he left off.

C. D. is not shy, but he's not really at ease either. And he's not ready to take me up on the rig yet. So I keep my notebook closed and we talk for a few hours until things feel a little bit normal.

Somehow we end up talking about sheep. C. D. and I both tended flocks of sheep as kids. Like me, he raised some lambs with bottled milk, which turned the lamb into a sort of woolly family member—a sheeperson or a persheep—whose emotional depth is hard for nonshepherds to grasp. If you're a lonely farm kid, a sheep can provide a great deal of nonjudgmental empathy. Sheep really know how to listen. Why beat around the bush? C. D. and I are both unreformed sheep-loving nerds. His pet lamb was named Rainbow, and sometime around midnight he begins to cry at the thought of the cruel uncle who killed Rainbow and ate him. C. D. first tries to wipe the tears from under his glasses and finally takes off the glasses and stands under the fluorescent lights, looking very gray. He puts his glasses back on and snorts, regaining the Texas macho thing. "I honestly thought about poisoning that son of a bitch," he says of his uncle. I think then that he is shockingly brave, but he prefers bravado. In any case, things feel normal enough to leave the trailer.

**Outside the trailer** there is a loud metallic shriek. C. D. harrumphs happily. "They're gonna trip!"

Tripping is the process of pulling up a worn-out drill bit, replacing it with a new one, and shoving the drill bit back into the drill hole. Sounds simple enough, but the hole is more than 10,000 feet deep, and the drill pipe needs to be removed from the hole in 90-foot sections, unscrewed, stored, and then reassembled in the same order with a new bit. That's 107 pipe sections, hanging in a sort of pipe rack on the side of the derrick.

C. D. leads me out of the trailer, past the three diesel locomotive engines that send power to the drill, past the red blowout protectors—giant shock absorbers that sit under the derrick and above the drill hole—past all manner of bulbous, ridiculously large equipment painted red, blue, and yellow. We head up the stairs leading to the "doghouse," which is the name for the white trailer hanging from the derrick forty feet off the ground.

There's an oil industry expression, "live by the drill bit," which means that the only real measure of a project's worth is the oil or gas that gets found. Intentions, promotions, hopes, and personalities fade in front of the mighty drill bit. From the drill bit America stretches out like a Rube Goldberg machine: Here is the bit that gets the gas that produces the electricity that powers the factory that builds the trucks that burn the diesel that . . . In this context, climbing the stairs to the doghouse has some aspects of a religious pilgrimage: Upward to the drill bit! Upward to the truth! On the other hand, the engine racket is so powerful the staircase shakes. The air has been supplanted by tangy diesel exhaust. Any sensible creature who isn't getting at least $13.25 an hour has long since left the vicinity.

The driller stands in the door of the doghouse, smiling down at us. He looks like a Doobie Brother, with a handlebar mustache, longish gray hair, and crinkly eyes. Between his accent and the noise, I understand about one word in eight. The doghouse is dingy—a sort of industrial clubhouse—with a bumper sticker for 1-800-BLOWOUT, a few lockers, and one of those indented aluminum stretchers used for immobilizing spinal injuries.

From the doghouse, you can step onto the rig floor, a large platform that surrounds the turning drill pipe. The driller controls the rig floor and above, and another foreman, called the tool pusher, controls the equipment on the ground. There is bright, rigid order on the rig, and each piece of equipment is painted a primary color. The machines are frequently hosed down, and now they glisten under the lights. The driller takes his place at the blue drill console, while a worker stands in the derrick over our heads and the three rig hands fidget around the drill pipe. The hands are all young: Jeans, overalls, fresh pink cheeks, and carelessly spread dirt lend them the innocence of characters in a Norman Rockwell painting.

C. D. has named them the Worm, the Midget, and Lost and Found. "Worm" is the name traditionally given to the newest guy on the rig. This Worm has been here only ten days. He has a square blond head, shoulders so large they appear to have football pads injected subcutaneously, and a constantly apprehensive, put-upon expression. The Midget is a wiry youth with an enormous weasily grin. Lost and Found exhibits a near-Buddhist detachment from the proceedings yet never screws up. It's unclear whether this is because he's got too many brain cells or too few.

The driller draws up the first length of pipe. The Midget leans from the waist, tossing a clamp (called "the slips") around the pipe at floor level. Without the slips, the entire string of pipe could fall back into the drill hole. C. D. launches into a monologue about what happens when things fall in the drill hole. I'll shorten the lecture: very bad and very expensive.

To unscrew one section of pipe from the next, the Worm and Lost and Found pull out two swinging contraptions called tongs and throw them at the drill pipe above and below the joint. The steel tongs are about four feet long and heavily and unevenly weighted. They swing in asymmetrical arcs close to the Midget's helmeted head, which makes me think of the stretcher, waiting inside the doghouse. The men stand back while the electric pulleys on the tongs unscrew the sections of pipe. Then they all lean in, detach the tongs, and push the pipe sections apart. The driller hoists the section to the top of the derrick, where the derrick hand grasps it, writes a number on it, and hangs it on a long arm protruding from the derrick. Already the driller is lifting the next section of pipe. The process continues, faster and faster, until it becomes hypnotic and beautiful.

"Tripping is like a ballet," says C. D. "Every guy has a position and a job." I disagree with him here: I think tripping is more like a square dance, an extremely fast, almost mathematically determined set of movements for five men; speedy repetition makes the movements seem almost giddy, but they are no less precise. Tripping is the apex of rig-think: Every action is purposeful, heavy things move fast, and dangers hang in tiny details.

C. D. leans in to point at the Midget's grimy shoulders. "The only way to work with the pipe is to use your latissimi dorsi," he says, poking at the Midget as if he's a horse. The Midget grins. "You can see who's working right by who's got the dirtiest shoulders," C. D. says, pointing at the Worm. The Worm's shoulders are still pink and clean, though the rest of his clothes and his face are smudged with grease. The Worm doesn't know how to pace himself yet, and has been vomiting from the heat.

Fear for the Worm: 42 percent of injuries happen to people who've worked on the rig less than three months, while less than 2 percent happen to men like the driller and C. D., who've been on the rig more than ten years. By the end of 2003, seven people will have died on U.S. land-based rigs. (In 2004, when many more rigs were at work, the number doubled to fifteen.) Injuries are common, and (according to the International Association of Drilling Contractors) most involve being "struck by," "struck against," and "caught between/in." Vulnerable body parts are, in descending order: fingers, backs, hands and wrists, feet and ankles, heads, legs.

Tripping, for all its weird industrial grace, is about speed and its corollary: money. A rig this size costs approximately $70,000 a day to rent and staff. At this depth, tripping the bit will take four hours—more if the crew moves slowly, less if they can keep themselves to a smooth unbroken rhythm. The shorter their tripping time, the more time the drill bit has to do its work, and the faster the well will be completed. There was a time when a well's profitability was determined by the potential bonanza beneath it, but developers of gas wells in this area know they're headed to a finite amount of gas, so they depend on reducing drilling costs to make profits. Between 2000 and 2002, the developer of this well cut the drilling time in this area from forty-seven days to thirty, reducing the cost per well by about $1 million. The company offers the crew an incentive to meet their target—a bonus of five days' pay.

The bit emerges from the hole two hours later. It's a frightening lump of metal: brassy, resembling the maw of a housefly magnified to the size of a football helmet. And it's steaming, because the temperature in the bottom of the hole is much higher than even this hot night. The crew gathers to appraise the bit, and they don't like what they see.

The bit consists of three cones tilted inward, each covered with medieval spikes. The cones rotate to crush and chisel at the rock while nozzles in the stem pump out lubricating mud at high pressures to wash away debris and make the bit turn more easily. This bit is prematurely worn down by nearly an inch all around, which is possibly why it lasted only 16.5 hours—half as long as it should have. If the crew has to trip another bit like this, they run the risk of losing their bonuses. The derrick hand comes down from his perch and stares at the bit in disgust. Bits are chosen by headquarters, explains the driller. "If the competition sent a gopher down the hole with a hardhat and a pickax, you'd see this outfit putting an ad in the paper: Gopher Wanted," he says.

Bits are so expensive—costing as much as a new compact car—they're kept in a locked cage outside of the trailer belonging to the company man, who oversees drilling. The Worm trudges up the stairs cradling a new bit in his arms, and in a minute it's down the hole. Tripping in reverse: The men's boots come near the pipe and back away. The tongs swing near the Midget's head—and miss—107 times. Body parts could be struck by, caught against, and caught between, but they're not. The pipe goes back down the hole.

If you want to understand the lean economics of modern oil and gas exploration, the bit is a good place to start. Tricone bits are not new—they were invented in the 1930s at the company founded by Howard Hughes's father—but the pace of development has speeded up as the exploration has gotten more competitive, committed to speed and margins. These bits use complex assortments of metals, diamond hard in some places and soft in others, to absorb shock. Nozzles control lubrication in ways that speed up drilling. "We think of the bit as a low-tech item," says Gary Flaharty of Baker Hughes, the biggest bit manufacturer, "but they're extremely high tech, with a development cycle of eighteen months—the same as computers." In 2003 one company drilling for gas in East Texas changed bit styles and reduced its drilling costs from $120 per foot to $89.

Hours later the driller flicks some switches on his console and tells me to pull a knob. By this time it's three in the morning and I'm getting loopy. The knob vaguely reminds me of a humiliating joy-buzzer party gag I experienced when I was seven. I hesitate to touch it, probably looking kind of slow and stupid. The driller looks expectantly at me. I pull the knob. Noises start but nothing in particular happens. "You're drilling," he says. Above us, near the crown of the derrick, a vigilant knuckle of metal begins to turn, holding 10,000 feet of drill pipe. "That's the Kelly joint," C. D. says with reverence. He begins an argument with the rig hands over whether the Kelly joint is the single most important piece of equipment on a rig, or not.

C. D. maintains the reason the bit wore down so fast is that the mud is all wrong. The unsung sidekick of the drill, mud lubricates the bit, stabilizes the walls of the hole, and prevents blowouts. On the other side of the drilling floor, the mud lies in big steaming vats the size of several municipal swimming pools. C. D. bounces across a woven wire floor suspended over the churning brown slurry. Fat pillows of steam rise from the mud, which has been heated by the high temperatures of the drill hole. The steam has a funny chemical edge to it, and I feel dizzy staring down into the whorls. Mud's name and appearance are deceptive: It's not like mud-pie mud, but a carefully calibrated and sometimes toxic mix of chemicals containing clay, barite, caustic, and mica—occasionally including mercury. Though drilling mud contaminates groundwater if allowed to escape from its tanks, the EPA has exempted it from hazardous materials regulations since the late 1970s. Texas, however, requires that companies apply for permits to dispose of the waste in approved ways.

People who live near drilling pay a high price for rogue mud. In 2004 oil and gas drilling had fouled the water in at least 241 sites in Texas with salt water, hydrocarbons, barium, mercury, chromium, hydrochloric acid, glycol, and PCBs. Local activists believe that number may be much higher, in part because the Railroad Commission (which regulates drilling) monitors groundwater quality at only 55 out of more than 4,000 disposal pits and because penalties for leaks are usually less than $3,500. "It would be fair to say that for decades, oil and gas drilling wastes have polluted ground water across vast areas of Texas," writes journalist Rusty Middleton, citing thousands of complaints of tainted water from eighty-five Texas counties. Near the town of DeBerry, for example, fifty-five people lost their water supply to benzene, oil waste, and salt water when a driller working nearby disposed of drilling wastes by forcing them into a well, an approved disposal method. Nine years after the first leaks killed peach trees and turned Earnestene Roberson's sink green and oily, the community is still without water.

**The only gushers** I see during my trip to Texas are the ones on the shot glasses I buy at the souvenir shops near the old oil fields. A gusher now is a blowout—an ecological and economic catastrophe. The gusher was always a bit of a scam: Promoters induced them with explosives on the theory that nothing parted an investor from his or her money faster than the sight of a plume of oil spurting out of the ground. The exuberant gamble of the gusher is the exact opposite of the firms working in East Texas today, which value control, steady return on investment, careful margins, and minimal waste of time and money.

After the oil output of Texas peaked in 1972, it was propped up by increased drilling when oil prices were high in the early 1980s but output dropped steadily after that. Even though many of the major oil companies, such as Texaco, Gulf, and Humble (which joined with Standard Oil of New Jersey to become Exxon), got their start in Texas, they quickly moved offshore and overseas, to places where the pickings were bigger. Texas's depleted landscape has been left to a swarm of small frugal independent oil and gas companies—bean counters, not wildcatters.

The new players depend on calculations rather than luck, which lost its influence when technology made finding oil and gas more predictable. Twenty years ago only one hole in ten yielded gas or oil. With seismic surveying and the ability to steer the drill bit more precisely, the odds are reversed. Now only one hole out of nine will be dry. And in this part of Texas, the company XTO claims that 97 percent of its holes produce gas. A presentation to potential investors uses the words "consistent," "low-risk inventory," and "highly efficient exploiter"—hardly the words to lure in the gamblers of the old days.

But the risk averse are happy with what they get: XTO said it got an 80 percent return on investment in 2004. One of the company's founders, CEO Bob Simpson, made $40 million for himself that year. But even he finds it necessary to attach himself to the old myths, telling reporters that he was attracted to the "romance" of the oil industry partly because he liked the smell of gasoline as a kid.

**C. D. carries** a coffee can of muck churned up by the drill bit back to the trailer. Inside, he spoons some of this dark glop into a sieve and begins to wash it. He tells a long extended joke about the marital problems of Minnie and Mickey Mouse, ending with the punch line "I told you she was fucking Goofy!" It's 5:00 A.M. I've been up nearly twenty-four hours. The lights in the trailer are getting brighter and dimmer on the same schedule as my pulse. "Your eyes look like tomatoes in milk," C. D. tells me.

Washed, the muck is made of cuttings that are hard edged and glittery. Under the microscope they gleam like jewels. It's been 100 million years since they've seen the light of day. I want to congratulate them, throw confetti, maybe try to interview them. Definitely time for bed. Possibly hallucinating. This is the closest I'll ever get to "seeing" what the bit sees, to touching the place where hydrocarbons are found. And yet I can't honestly say it looks different from ordinary sand. In the center of the cuttings sits an attractive pale greenish clump. C. D. hurrumphs and says cryptically, "Carboniferous. Glauconitic." When I don't respond he makes a honking noise. "Pond scum!" he crows. Deflated, I leave to get some sleep.


**Pond scum.** Oil and gas come from pond scum or, more commonly, ocean scum: phytoplankton—little single-cell chemical powerhouses. (Other possible ingredients for oil include leaves and fecal matter.) The gallon of gasoline I burned driving to and from the rig ran me up a planetary bill of approximately 89 metric tons of green scum.

Phytoplankton lead short lives of great consequence: Within forty-eight hours they're born, photosynthesize, and die. They drift to the bottom of the sea holding carbon dioxide within themselves—enough that marine sediments made of phytoplankton corpses now tie up 99.9 percent of all the carbon dioxide that's ever been part of living things in all of geologic time. Yes, it's an absurd statistic, but without phytoplankton, there would be no life here at all.

And there would be no oil. There's a common misconception that oil and gas come from dinosaurs—or "400-pound oysters, 300-pound starfish, and clams three feet across," as one Texan in the oil industry told me. Perhaps it's a slight, or a conspiracy, but phytoplankton don't get much respect. Without the phytoplankton, there would be no Houston Astrodome, no Dallas Cowboys' cheerleaders, no NASCAR, no Space Center to talk to the moon landings. And chicken fried steak never would have evolved. It's time for the phytoplankton to take a bow. Texas likes to reward its heroes with statuary. A sixty-six-foot-tall statue of Sam Houston is visible for six miles around the town of Huntsville. I think a seventy-foot-tall statue of the mighty phytoplankton of the Cretaceous would be a fitting tribute.

Back to the oil and the gas. Hydrocarbons are constructed by serendipity, a series of productive accidents. A recipe for making oil starts with a nice warm ocean basin with edges that encourage sedimentation. Phosphorus and nitrates nourish the phytoplankton until they're reproducing with gusto. Corpses drift gently to the bottom. And between 90 and 99 percent of the little carcasses will get consumed by bacteria. Game over.

But if the basin happens to have an oxygen-starved bottom, a few percent of the phytoplankton will decompose until they're mostly fats—carbon, hydrogen, and oxygen—mixed in with silt. Give the stuff a million years, as sediments pile on top of it, and the fats will form kerogen—a brown or yellow-brown solid that is chemically different from the organic stuff it came from. Kerogen is the intermediate step between scum and oil, much as dough is the midpoint between flour and bread.

To find out how kerogen actually turns into oil, I called Dr. Michael Lewan of the U.S. Geological Survey in Colorado, who came up with a way to cook oil from kerogen-rich shale in a laboratory in 1978. "As simple as it is," he said, "you'd think someone would have done it before then." Lewan starts with some uncooked shale, crushes it into gravel-size bits, puts it into a stainless steel cylinder, covers it with water, and bolts the cylinder shut. Then he drops the cylinder into a heating jacket and turns the temperature up to somewhere between 300 and 365 degrees C (572 to 689 degrees F). Seventy-two hours later, he removes the cylinder from the oven. Voilá, crude oil floats on top of the water.

But if you can make crude oil that quickly, why does it usually take eons? Think of Lewan's cylinder as a time machine: By raising the temperature, Lewan is able to condense hundreds of thousands—maybe millions—of years of geologic time into a few days. "People always ask me why I don't just drop the temperature to 250 degrees [482°F]," he says, "but then I'd have to run the experiment for one hundred years. And I know my kids are not going to do these tests for me when I'm gone."

In real geologic time, Lewen's oven is underground, which gets hotter the deeper you go. (For every kilometer of depth, the temperature increases by approximately 25 degrees C. For every mile, the change is 72 degrees F.) If this kerogen makes it to a kilometer or two deep, where the temperature is at least 60 degrees C (roughly the internal temperature of a convenience store hotdog) the process will start, but it will take a long time. If it gets pushed deeper, to around 80 degrees C (roughly the temperature of a scalding cup of coffee), the oil will form faster.

Lewan has been able to observe the process by yanking his cylinder out of the oven early. The heat works somewhat like a refinery in reverse: Instead of breaking the hydrocarbon chains apart, the underground temperature forms the hydrocarbons into long strings. Lewan has found that kerogen gets soft, arranging itself into long chains called bitumen. The bitumen changes its shape in the rock, and—almost like melting butter in pie dough—forms little interconnected networks to join up with other bitumen. Under further heat, the hydrocarbon strings in the bitumen crack into a stew of smaller molecules, some liquid and some gas, to make crude oil. But these smaller molecules take up more space than the long bitumen chains in their old rock home. The crude oil, now cramped for space, is under pressure. Sometimes it will use the bitumen networks to flee the rocks where it was formed. If the crude happens to find nearby rock that's porous—say a sandstone with extra space between sand grains—the oil will migrate there. It will keep migrating, floating on top of water until it finds something it cannot pass through—shale or salt, or a fault that has placed a layer of hard, impermeable rock at the end of a porous layer. And there the oil will sit, trapped under pressure, until some bit comes along and finds it.

Unless . . . If the oil is pushed still deeper into the earth—below 16,000 feet or over 145 degrees C (290 degrees F, less than the temperature of the fat used to fry doughnuts), the heat usually acts like a refinery—cracking the crude oil into lightweight molecules of natural gas: methane, butane, propane, and the deadly hydrogen sulfide. And then that gas sits, trapped down there somewhere, under great pressure. Waiting.

### Rig: Night Two

The Fifth Avenue is in the same place, hatchet still resting on the engine block. Hanging in the window is a T-shirt reading: GIVE BLOOD: PLAY HOCKEY. The drill bit is at 10,137 feet, which is 109 feet deeper than last night.

Inside the trailer, C. D. is examining the logs—accordions of graph paper covered with impenetrable squiggly lines. "We've got maybe 453 feet of lower Cretaceous left before we cut into the Jurassic," he says, as if this were no different from driving from one town to another on a highway lined by strip malls. Under C. D.'s influence, I start to adopt his attitude that there is predictability and repetition in things both below- and aboveground. The numbers, as he would say. I suspect that C. D. has bestowed the names Worm, Midget, and Lost and Found on roughnecks on many other rigs. Shale, rig hands, blowouts: Every grain of sand is a chunk of some bigger, known rock, part of a story older than the hills.

Oil is slippery but also predictable. Sometimes oil will move a hundred miles from the shale where it's created to the sandstone in which it's found. So if you're a geologist or a mud logger, you need to be able to psych out the oil—anticipating its movements through an underground landscape. In a sense, searching for hydrocarbons is like following a feckless hitchhiker—oil hitches rides on water, usually, and takes the path of least resistance until something prevents it from moving. Then it sits passively, pressure building, until the bit finds it. If you can find where the water is going and where it stops, you may find hydrocarbons too.

C. D. spreads out the wireline logs—grids containing squiggly lines—and explains how he uses them to "see" what's underground. The logs are the result of sticking electric probes in the well hole, putting out different currents, and measuring the feedback. By looking at the relationships between the different squiggles, he is able to see the qualities of the rock layer. He points out a furious squiggle on the chart—evidence of a lot of gamma rays, a sign of shale. C. D. says this is the Rodessa shale. He turns to the logs again, showing that the other two log lines in the Rodessa are flat, evidence that the rock is not porous or permeable, which means it's not holding extractable oil or gas.

"We're down in the Travis Peak formation, looking for that sandstone," C. D. mutters. Travis Peak, he explains, contains a lot of pink sandstone, remnants of a huge red mountain in the middle of the Cretaceous Sea that weathered, spreading particles of its pink self across the seabed. He shows me a sample of the sand—it's got fine grains, half white and half a rusty red. With a powerful microscope, you can see the bug fossils in the sandstone.

A geologic map of Texas looks roughly like a steak. In the center right, like a ribeye, sits a red circle. That is the remains of the giant red mountain—the Llano Uplift. C. D., still flipping through his accordion-pleated logs, tells the story of how he learned of the Llano Uplift. "One day my grandpa was driving from Odessa to Snyder and we stopped in a place called Tarzan to get what my grandpa called a soda water. When we got back in the car he said, 'Close your eyes. I'm going to take you on a trip on the Cretaceous Sea.' He told me to imagine that we were in a boat and we could see a big red mountain sticking out of it. When I got in the boat, he said, it looked like it would take an hour to reach the mountain, but the mountain was really so large it took days to get there. He said it was Cretaceous Texas—between Llano and Burnett. It was like we were traveling underground. We were way down in the Hosston formation."

As soon as he's finished with this story, C. D. seems slightly embarrassed by it. He says we need to go catch a sample. Outside, the insects are still hopping up and down. The Worm is vigorously scrubbing the staircase. I turn toward the southwest, trying to imagine that big red mountain.


**Feeling some** deep connection to the underground and the prehistoric is an attribute of this fraternity of oil. In Houston I met a geologist who described his work as "a novel" and then went on to describe an obsession that sounded more like an epic quest than a career with a major oil company. "I've glimpsed great mountains beneath the sea that no one ever sees," he said, glancing around his spacious Houston home. "Being able to interact with the subsurface data is the drug that keeps me here."

The geologist requested that I not use his name and then offered to show me the novel. We sat in his rec room while he accessed his company's mainframe computer with his laptop. His house was calm and beige, several languid rescued greyhounds snoozing at our feet. The geologist began to manipulate seismic maps of a landscape five miles deep. The maps were made by creating small explosions on the seafloor and then mapping the ways the sound waves bounce between strata belowground. Seismic maps are not only 3-D, many also incorporate a fourth dimension—time—that is rendered visually through algorithms. The images on the laptop are beautiful and subtly colored, and in the atmosphere of his living room, they remind me alternately of raked gravel in a Zen garden and slices of malachite. By changing the way the mathematical data is processed and using different colors, he is able to "look" into the formation in different ways. One minute the image looks like marbleized paper—violent swirls of color—and with a quick change of the math it resembles hammered aluminum. He turns the strata so he can see a formation the size of a Volkswagen. As he manipulates the images, the geologist loses his reserve and begins to yell at the mainframe to go faster.

Here's the story he tells: Once upon 150 million years ago there were sands on the edge of the continent. There was a storm. "The sands come screaming down the edge, with the small sand spilling like water and the bigger like molasses. Big rocks are coming over the toppings like a giant Pachinko machine." When the debris landed, it was sorted perfectly by size. And then it was sealed in by shale. The hydrocarbons cooked in the shale and, over the course of a million years, they migrated into the sand. "Presumably the oil migrated through a hole right here," he says, pointing at the screen. And there were the faults and folds, with individual histories as compelling as subplots in Russian literature. And now, inside that trapped sand, sit water, oil, and gas. "It's a historical science, like reconstructing the scene of an accident. It's very creative, addictive, fascinating," he says with an air of helplessness, as if his love is beyond him.

And then he comes back to his senses. "Of course, the whole goal here is to map the living shit out of this to mitigate the risk of drilling a $150 million well."

**Harrumph.** In the trailer, C. D. looks at a sample and says, "Put a dress on that pig and take it to the fair." Under the microscope, the cuttings are mostly a bed of white sand. "See that rock, floating in the middle like a pale green ankle sock with its toe pointing that way?" I do; it's a small, greasy, unimpressive chip. C. D. says it's glauconitic shale, which means we're leaving the sandstone and beginning to approach the next layer of shale.

He's waving an eyedropper over the sand, explaining that if the acid fizzes, he can tell that the sand grains are cemented together by calcium. (The other kind of cement is salt, which won't react, but it will dissolve in water.) Calcified sand can act like a cage around trapped hydrocarbons. When the bit drills into the cemented sandstone, only a small amount of gas or oil will be able to make it to the well hole. The rest sits trapped until the cemented walls are broken by a combination of explosives and high-pressure water.

The gas we're headed to is in cemented sandstone called Bossier Sands. The Bossier are the remains of pinnacle reefs—limestone reefs covered with sand deposited by rivers at the edges of the Cretaceous Sea. When the rivers changed course and stopped dropping sand on the reefs, the reefs were covered by shale. If you were to cut the area apart with a knife, you'd have something that looks like a chocolate bar with almonds in it: The reefs would be the almonds and the chocolate coating would be shale. (The shale below the Bossier Sands is called the Smackover.) Oil was formed in the shale, and then it migrated into the sandy reefs, which had spaces between the sand grains. When the whole chocolate bar was jammed to deeper, hotter depths, the oil that was hiding in the "almonds" of sandstone cracked and became gas. The pressure grew inside the reef, but the sand itself cemented, trapping the gas in its pores. And then it waited.

The remarkable thing about the Bossier gas is that it didn't officially exist until 1996 or so. People knew the gas was there, of course. But the Bossier had been considered nuisance gas—its high-pressure pockets threatened blowouts for rigs drilling to deeper, more profitable gas. A 1995 survey of gas reserves didn't even list Bossier, which is estimated to hold 5 trillion cubic feet of gas—enough to make electricity to provide three years' worth of air-conditioning to the entire United States. In its sandstone cage, Bossier belonged to the large group of oil and gas reserves that cost more to get out of the ground than they're worth above it. So the Bossier gas sat underground, unwanted and unremarked—until the boom commenced.

**If it weren't** for the fact that all of the motels are full, I wouldn't know that Fairfield is a boomtown. There are none of the gambling parlors and honkytonks and women dressed in "beach pajamas" that characterized boom days of the Black Giant in the 1930s. There are, however, men in pajamas. In the center of town, in front of the square old courthouse, the air has a thick, syrupy feel. The only people moving around are two men wearing striped pajamas, tending the shrubs in front of the courthouse in slow motion. It takes me a Texas minute to realize that they look like the man on the monopoly board because they're prisoners. Orange jumpsuits haven't made it here yet. Anyway, the courthouse is the epicenter of Freestone County's gas boom.

Before a company drills a well, it needs to secure the permission of the owners of the mineral rights for that property. In Texas, surface rights are often sold separately from mineral rights, which means that finding the owners of the mineral rights can take a lot of digging in old court records. Exploration companies hire agents called landmen to do the research and get the contracts signed. As the boom in drilling heated up in Freestone, more and more landmen came to the courthouse. Between 2003 and 2004, the *Wall Street Journal* reported that the Freestone courthouse photocopy machine's annual revenue increased 30 percent—to $99,000.

Across from the courthouse there's a landman's office. Through the window, the owner looks a bit like Kris Kristofferson. LANDMAN is painted in gilt on the window. I am headed toward the door when I notice that he's admiring a rifle, and that seems like a private moment. So I call him later.

Texas laws are set up to favor drilling, so if the mineral rights owners sign the papers to get the royalties—an eighth of the production of the well—the surface owners can't stop them. The landman has hunted down mineral rights owners in places as far away as California and Chicago. "It's real hard to keep a driller out of a mineral," he says.

What that means, in practice, is that Texans can't control what's in their backyards. It's one of the ironies of life here. The fields around Fairfield are dotted with Christmas trees—the stubby pipes protruding from finished wells—and pumpjacks, which are the big nodding black pumps that pull oil to the surface. There are pumpjacks whirring and clacking in town centers, beside schools, and in idyllic farmland, surrounded by horses. The landman says his brother bought a ranch and soon it was dotted with seventy-five to eighty wells, "not what he wanted." The wells are either beautiful or monstrous, depending on your perspective, but the royalties from them go to the owners of the subsurface rights, not to the landowner who has to look at them every day.[1]

Still, oil and gas royalties mean there's money floating around. In 2002 Freestone residents declared $39 million in royalty payments on their income tax, which works out to more than $6,000 per person. In

---

[1] Owners of surface rights do get compensation for the hassle of the drilling, but they don't share in royalties. The landman has to pay them for lost livestock. "I get a call in the morning that a cow has fallen into the mud tank. It never fails, it's always a prized bull. I show up with a checkbook and say, 'Now, how much was that prize bull worth?' " There is a long history of cows coming to a bad end near oil and gas drilling. A few: Thirteen of twenty-eight cows were found dead near a gas well. Their postmortem blood was chocolate brown. Fifteen cows were found dead near an oil rig after they'd been observed licking a white granular material from fifty-pound sacks.

this atmosphere, the landman says, deals are easy to make. "People say anything I've got is for sale except the wife and dog."

In the 1920s, he says, there was a rush in oil, which was found in shallower, cheaper-to-access deposits. "They're just continually drilling deeper," he says. "It's 13,000 feet deep now, and some are even 20,000 deep. In ten years it'll be even deeper." The boom now is for gas that was too risky with better prospects available elsewhere. "Ten years ago you couldn't do this without being lucky."

To find out what had made a risky gas pocket into a boom-worthy one, I called the Texas Bureau of Economic Geology at the University of Texas in Austin. Their resident expert on natural gas is Dr. Eugene Kim, who doesn't mention luck: Fairfield's boom is driven by government politics and policies.

Today, he says, most of the "easy" gas and oil has already been gotten from the United States. What remains in big quantities is "unconventional" gas—the methane that occurs with some coal deposits or "tight" gas, like the Bossier gas, locked into its cage of cemented sandstone. In the late 1980s the federal government and the gas industry put up $165 million to figure out how to get tight gas out of the ground. They ended up finding ways to fracture reservoir rock using explosives and pressurized water deep in the wells. Then, to encourage more drilling for unconventional gas, federal and state governments lifted taxes on it. Worthless gas became profitable, and a boom slowly began. The boom in Fairfield is more and more typical for the country because now "unconventional" gas contributes a third of U.S. production; within fifteen years it will make up half. As natural gas prices have risen, unconventional gas looks profitable. "A lot of our future resources are coming from tough areas," says Dr. Kim. "That requires new technology and state and federal incentives."

When oil is extracted from a reservoir, between 50 and 75 percent remains in the ground. It's likely that better technology would improve that. The International Energy Agency estimates that increasing the world oil recovery rate from 35 percent to 40 percent would result in more "new" oil than all of Saudi Arabia's current fields. Even though studies by the National Academy of Science have shown that oil and gas technology research is a worthwhile investment in future oil supplies, money for it has fallen since the mid-1980s. Major oil companies have abandoned their expensive research programs—partly because oil prices were very low in the 1990s and partly because the industry restructured to focus on profits. The small companies, driven by bottom-line economics, haven't taken over research. In the meantime, the Department of Energy has invested less in research. In 2005 the Bush administration tried to do away with it altogether. Dr. Kim (whose workplace benefits from government funding) says that politicians are reluctant to spend money on fossil fuel research because it looks like they're offering freebies to the energy industry.

Tax incentives, however, often slip below the political radar. One calculation estimates that Texas gives $3.5 billion in tax breaks to oil and gas producers, which supporters claim pays for itself by generating twenty-two times that amount in income from oil exploration and production. These estimates are self-serving, but they do reflect the reality that many people in Texas depend on oil and gas, in one way or another, for their livelihood and income. Another industry group estimates that all U.S. state incentives for oil and gas exploration total $5.5 billion per year but bring in twenty-eight times that in indirect economic benefits.

"It behooves everyone to understand the cost of 'cheap' energy," says an enthusiastic report on incentives by the Interstate Oil and Gas Compact Commission, and that's really the problem with tax incentives: Although the incentives shield consumers from knowing the true cost of oil and natural gas, they still have to pay. An entrenched system of tax breaks for the energy industry, which include allowing drilling on federal land, relaxing royalty payments, and allowing companies to write off certain expenses, adds up to many billions of lost tax dollars yearly. One federal program, for example, will give companies access to $65 billion in oil and excuse them from paying even $7 billion in royalties. And a series of lawsuits filed in Colorado accuses three hundred companies of defrauding the government of more than $30 billion. These enormous giveaways are not reflected in prices at the gas pump (or, for natural gas, in electricity bills).

But whatever I pay for energy, I'll never pay as much as residents of Texas do. It's Texans who give up their backyards to noisy drilling rigs, their water aquifers to possible contamination by drilling; who breathe the air near the refineries of the Gulf Coast. When oil prices fall and people are thrown out of work, they're usually Texans. And whatever benefits the average Texan has received from the state's romance with oil, they've sacrificed mightily and, like the owners of surface rights, without much choice in the matter.

Kim, who grew up in Korea before moving to Texas, is annoyed by what he sees as the chauvinist politics of oil in the United States, where Florida and California have banned drilling for oil on their coasts. "It's basically a NIMBY approach," he says. "We don't want to explore for more oil and gas resources yet we want to consume them."

## Rig: Night Three

The bit is at 10,457 feet. The Fifth Avenue still has its hood up, though the hatchet is gone. Inside the trailer, C. D. looks ill. And in the parking lot, there's a woman sitting in an SUV, staring straight ahead at the rig. My gut sense, from the rigid angle of her neck, is that she's either very angry or a bit crazy.

Back inside, C. D. is trying to make the logs line up with the progress of the bit. He suspects that the daytime mud logger has been distracted. "Bill's not himself," C. D. says reprovingly. "He's reading those romance novels and he's not even sweeping the floor. Bill's not like that."

I'd met Bill, a cheerful former marine in his fifties, and I hadn't figured him for a romance-novel type, whatever that is. Bill looks like a cartoon drawn with one of those if-you-can-draw-an-oval-you-can-draw-this-dog books advertised in the back of the comics. He has a shaved oval head with sunny oval cheeks, a sideways oval chin, and mirrored bubble safety glasses. Top that with a hard hat, attach it to an egg of a belly and bulbous steel-toed boots, and you have Bill. A devout Christian who lives in an RV with his wife and son, Bill broadcasts good-natured goodwill in all directions. At the moment, though, he and his wife are absorbed in a series of nine romance novels by Janet Dailey. Bill apparently spent the day scouring the Internet for Dailey's e-mail address to ask her a question about a detail in one of the books.

I offer to catch a sample of drill cuttings and eagerly sprint up to the doghouse, say hello to the driller, the Worm, the Midget, and Lost and Found, and collect the sample.

When I come back down the stairs, the woman in the SUV has moved. She is now at the back of the rig. She turns on her high beams but the light gets lost in the halogen flood at the base of the derrick.

In the trailer, C. D. washes the sample and puts it under the microscope. The sand is a rich Llano Pink and I'm back in the car with grandpa, floating toward the giant red mountain while little C. D. sips his soda pop through a straw. But among the grains something sparkles like tinsel. It's gold, fitting too neatly into the magical scheme of the underearth I've been nurturing this week. C. D. looks into the microscope, grabs a magnet, and waves it over the sample. "That's bit!" he exclaims, pointing out the crumbs of worn metal. "We're gonna trip."

A worker known as the derrick hand knocks on the door. A Louisianan in his early twenties, his brown eyes give him an expression that can be interpreted as deep sincerity. "Sheee-itt," he greets us. His accent is so thick his tongue seems to be wrapped in bacon, and every syllable has three beats. "It's hotter than a whole FAMily of mice in a WOOL sock," he says. He grew up in Louisiana's cotton country but quickly realized there was no money there. So he went into oil, which gave him a ticket out of Louisiana and into the international brotherhood of oil guys. Now he and C. D. swap stories about offshore rigs they've worked on. The food, he says, is wonderful. You can get a steak or shrimp anytime you want. But if a cook is bad, the rig hands make his life hell. On one rig, a surly cook was tossed off into the ocean from a height of sixty-five feet. "I believe I'd die before I hit the water."

C. D. inquires about the woman in the SUV. Oh, the derrick hand met her in town and she must have taken things wrong, because now she's moved her car so she can see him on his perch at the top of the derrick. Angry. Definitely angry. The driller told him to get her off the grounds, so he has come to the trailer to procrastinate. C. D. listens sympathetically, his man-motherly side dominating the smartass side for the duration. He doesn't emit a single sarcastic harrumph.

The derrick hand has other problems. His baby's mama in Louisiana has called to say she's sick from taking too many diet pills. "She thinks she's fat. She's not. She's very beautiful. But I'm away a lot," he says, shrugging. "Shee-it." This time he says it mournfully, even introspectively, as if his loneliness is collapsing inward. The roughnecks work fourteen days of twelve-hour shifts, get a week off, and then come back to work for another fourteen days. It's an alienating life, and the pay averages $13 an hour, though the derrick hand probably makes more. Anyone who can find a reason to quit does. He appears lost in existential worry.

Then his head snaps up, and he looks brightly at me. "What motel did you say you're at?" C. D. snorts and growls at him, "Gonna ask for her room number?" The derrick hand leaves to talk to the woman in the SUV.

When the door shuts, C. D. becomes dejected. "This is God's retard colony. We're the dumbass colony."

The U.S. domestic oil industry is dying. Seventy percent of U.S. oil workers have been laid off since 1981. It's an industry where insecurity has become the norm. Even people at the top of the food chain are at risk: The Houston geophysicist has survived sixteen layoffs, and he's only in his late forties. "It's nerve-wracking and you feel awful," he says, "but they're really just trading people around to reduce salaries and pensions. It's like selling equipment." People lower on the totem pole—the mud loggers, tool pushers, and even the drillers—have retrained after cycles of boom and bust. Now nearly half of the industry is between fifty and sixty years old. With a boom going on, most rigs are like this one—filled with worms and old old hands.

C. D. survived the busts in the industry by working with computers. In the late 1990s he formed his own start-up company to write an Internet program that would send mud log information directly to well investors, so they could make decisions quickly. His program was one of several that aimed to eventually replace human mud loggers with instruments and computers. But the mud loggers are safe for a little while, at least, because C. D.'s project got mired in legal troubles.

And that was a good thing, in a way, because when he needed to cover the legal bills, the first job available was mud logging. "I'm the red-headed stepchild, the wallflower," he says, with a look of grievous defeat, "a genetic deformity." Soon after, his kidney problems—the result of a bout with hepatitis C—kicked in, and he needed the health insurance. "Roper's law is that Murphy was a drunk optimist," he says flatly.

**The next morning** I drive to see the Black Giant in the town of Kilgore. It's a hundred-mile drive through small hot towns with greasy pumpjacks nodding in every clearing and pasture. Downtown Kilgore used to be full of derricks, but now it has a whooshing, cottony sound, as if it's a soundstage for a movie about the 1930s. Old buildings with colorful tile sit closed and blank. A metal sign advertises FURNITURE TIRES with no facilities for delivering either. There are forty-one derricks in downtown Kilgore today, and all but one are small bogus replicas, decorated with Christmas lights. A lone pumpjack whirs and screeches near what used to be "the world's richest acre." If oil fields could echo, this one would have the sound of 6 billion barrels gone.

The Black Giant is empty, but we still live in the world it created. It was so big, and gone so quickly, that it reset the world's center of gravity. As soon as the forty-two-mile-long field started producing, drillers descended, each trying to get the oil out of the ground and sell it as quickly as possible. With the glut, prices fell from $1.10 a barrel to 11 cents, and then to 2 cents. (Comparison: A trip to a Kilgore latrine was going for 10 cents; a hot meal or a shower for 35 cents; and women were selling sex for 50 cents.) It was classic oil economics—a bottomless supply means oil has no value. What looked like a glut was in fact a slow-motion bankruptcy, because each barrel cost 80 cents to get out of the ground. Worse, the oil was being drawn from the reservoir faster than the water that drove it upward toward the surface was being replaced. The field began to produce less oil and was clearly lurching toward ruin. A disaster of abundance was in the works.

One solution was to limit oil production to stabilize both the price and the pressure in the reservoir, but that was un-American, uncapitalist, and utterly anti-oil as we knew it. It took Texas Rangers, the National Guard, and a Supreme Court ruling finally to determine that shutting in production (or pro-rationing) was something the state Railroad Commission had the power to enforce. The amount of oil coming out of the Black Giant was limited by the government, and prices stabilized. The wildcatter had been brought to heel, and some claimed the independent oil producers became a kinder, less venal lot when they were forced to cooperate. I don't know about that.

The Black Giant started a backlash against waste. Until the 1930s, oil fields disposed of natural gas buy burning it in flares. Now they began capturing the gas and selling it as fuel, creating the natural gas industry.

Secretary of the Interior Harold Ickes took one lesson from the Giant's rapid depletion: "Without oil, American civilization as we know it could not exist." And so in 1943 he nudged President Roosevelt to think of oil and build a "special relationship" with Saudi Arabia, which had recently discovered the world's largest oil reservoirs. And with that move, the United States insured it would have access to huge quantities of cheap oil for as long as the relationship lasted. For all practical purposes, the idea of American energy independence was dead. The myth of the gusher and the wildcatter, though, lives on.

The Black Giant continued to change the world around it even after it faded from view. During the 1950s, a Venezuelan exile named Juan Pablo Pérez Alfonzo began reading the story of the giant oil field at the Library of Congress. He saw that cooperatively limiting production would give oil producers like Venezuela, Saudi Arabia, and Iran more power over prices while discouraging waste. From the Black Giant, he came up with the ideas that started the Organization of Petroleum Exporting Countries, OPEC.

## Rig: Day Four

Now the bit is at 10,822 feet. The hood of the Fifth Avenue is closed.

C. D. looks terrible. His skin is the color of putty, and his eyes do not focus. He has a gash on his arm covered with gauze. His anemia, a side effect of the kidney condition, is bad. No sooner than he admits this, he's talking even faster—a whole cascade of memories and dirty jokes, geologic observations, and miscellany. The Fifth Avenue's spark plugs are lying on the counter. If you clear away the bluster, he may be here on this rig because physically he cannot leave.

I leave the trailer and sit outside in the field with the bugs. Maybe he'll feel better if he doesn't have to talk. I don't know what I expected to learn when I came to Texas, but it was simpler than this. Perhaps I was hoping for a cartoon, a myth, a happy-go-lucky wildcatter. What I got was C. D., whose ferocious smarts and attraction to oil has led him away from safer jobs and comfort.

Added to that, I've gained a confusing Texas parallax view: The night sky forms a royal blue dome over the glowing rig. The sounds of crickets and diesel engines run in stereo like a dance beat. The Worm, the Midget, and Lost and Found run down the rig stairs like ants in the glorious anthill of industry. Like R. L. Gaston, for a moment I cannot see anything ugly about a drilling rig.

And I've only been in Texas for a week.

Back inside the trailer, C. D. is composed. He's calculated that we've got an hour or two of simple drilling in hard sandstone before the bit hits a pocket of gas, runs quickly for a few feet, and then zips back into hard sandstone.

He says neither of his kids has any interest in being fifth-generation Ropers in the oil business. "Oh, they love knowing about it," he says, "but I seriously doubt they'll do it." Even C. D.'s beginning to reconsider.

He tells a story about a blowout he saw in 1979, south of Corpus Christi. The crew was drilling in sand and shale with dangerous pressures. "The well took a kick," C. D. says. When the crew pulled out the drill pipe a vacuum was created, "like a cow pulling its foot out of the mud." He says the vacuum accelerated when wind blowing across the top of the pipe acted like the bulb on an atomizer. As the sand started gassing, the vacuum did the work of pulling the natural gas deep in the hole to the surface. There were no blowout protectors and the mud was not calibrated. Soon the violent strength of the escaping gas tossed the drill pipe into the air so it landed "like spaghetti in a bowl." The gas was on fire in an instant, joined by salt water and poisonous hydrogen sulfide. Soon the top of the well collapsed in on itself, becoming a great hellish crater vomiting poison, fire, and hot salt water. And it was growing.

That's when C. D. got the call. He looked out his bedroom window in Corpus Christi and saw a fire that looked as close as trash burning in the alley. He spent six days with another rig in a schoolyard near the blowout, trying to drill a new hole into the reservoir to relieve the pressure down below. "It was a blur. I slept an hour in six days. My cheeks were as red as suspenders because you'd step out the trailer and it was like being sandblasted. It made the face shields on our helmets opaque.

"The whole crater filled with the salt water from the well and became a great salt lake. There were butane bottles that had fallen off the burning trailers floating around in the crater. The bottles became round from the heat, and people would just kick them as they floated by in the brine. It was like everyone became John Wayne.

"What sticks with me was that if I stepped off the trailer, I had to be careful not to step on the merry-go-round because the relief well was in a schoolyard."

He stopped here to let the story sink in. What he's really talking about is a seduction, a legendary love story between him and the oil business, the chance to become John Wayne. Looking back, perhaps it's starting to seem as wonderful and ridiculous as any other love story, so he tries to explain it. "If you're in a blowout, your farts have lumps in 'em," he begins. "A big adrenaline rush. Everything else is boring. I'm realizing that I'm addicted to adrenaline. But now I'm in my mid-fifties and I'm thinking about this for the first time." His voice crumbles to a stop. We sit for a few seconds. I'm thinking about his kidneys. I don't know what he's thinking about. "You know, fear smells pretty good," he says.

**I wanted** to meet one of the old-time wildcatters, so I headed to Houston to meet Michel Halbouty. "They refer to me as the greatest wildcatter of all time," he says. The air-conditioning is so cold in his office that he wears a double-breasted cardigan, if there is such a thing. At ninety-four, his eyes glitter like drill bits behind his thick glasses, while the rest of him is a reminder of Houston's heyday: a precisely trimmed mustache, the drawling, nearly wrathful voice. On the office wall: a photograph of Halbouty in a tuxedo and his wife in a frilly ball gown waltzing off a well-head in one of his oil fields. This is the last outpost of sequins and Cadillacs in a town that's moved on to revolving layoffs and fortunes made through accounting.

Even Mary, his assistant of forty-six years, seems to have been dispatched from Central Casting half a century ago. She stands up to greet me—six feet of perfect grooming and Joan Crawford eyebrows—apologizing for forgetting my appointment by hollering, *hollering*, "Fire me now or fire me later!"

Halbouty sits me down near his massive burl desk. A stuffed collie in a Texas A&M sweater looks glassily at me. Son of Lebanese immigrants, Halbouty lived for oil, losing two whole fortunes and making yet another. "At one time I had fourteen straight discoveries followed by thirty-six straight dry holes," he crows. Grand old times. Why didn't he just retire? "Because that's me," he says, with outrage at the mere idea of being too comfortable to take a risk. At this very moment he has four wildcat rigs drilling for oil. He believes the United States is riddled with unfound oil, but people are just too complacent to try to find it.

A phone rings. "Excuse me," he says, "that's my private line. It's my wife." His voice changes to something pulled out of a honey jar, dripping. "Hi, honey. Yeah, baby." Maybe she always wears the ball gown.

He sits back down. "Reagan and I used to talk a lot. I was his energy guru." Halbouty is an oil omnivore—a geologist who's published more than three hundred papers. He is also a frenetic wildcatter, international oil diplomat, and the coauthor of a fantastic history of the Black Giant, where he started his career. He advised every Republican president before Bush II and supervised Reagan's transition team on energy by flying to Washington in his personal jet.

He is talking with me for one reason: The grand times are gone, and he is not optimistic. "I have lived in the most important time in the history of the United States," he says, poking the table between us for emphasis. "When we had oil. But this country hasn't taken care of the one commodity which means its life: energy."

Halbouty has been harping on the country's lack of an energy policy for more than thirty years. That alone says something about his persistence and about national disinterest in the topic. In his estimation, the country's position changed dramatically when we stopped producing all the oil we needed. We began a slow slide into "deindustrialization" and destruction. Without our realizing it, the pillars of our well-being began to erode. "Every empire in the world died after two hundred years, and we are destroying ourselves by not looking after our welfare and our welfare is energy!"

Halbouty ends with a high bark and looks at me indignantly. "If you don't get that in your book, you're missing the whole point."

He goes to get me a copy of a biography of his later life, titled *War Without End*. In it he advocates conservation of energy, more oil exploration on federal lands, more tax exemptions for energy exploration, more investment in traditional and alternative energy research, more coal and nuclear power, and cheaper environmental safeguards. Some of these are standard boilerplate requests from the energy industry, with the twist that Halbouty wants policies to favor independent wildcatters rather than major oil companies. And some, such as conservation and more research, are long-overdue elements of a national energy strategy.

"I don't care about me," he says. "I got everything I want. But this country needs an energy policy or we just gotta take all the oil we can get and say the hell with it."

"Say the hell with it." Conjure end times: Houston's freeways are flat mesas with scattered packs of ragged desperados searching for gasoline.

Cold winds blow. But as soon as I exit Halbouty's reality, it will be 107 degrees outside and the freeways of Houston's sprawl will be jammed with cars and trucks, as usual, regardless of the hour. The slightly humorous thing about Michel Halbouty's dystopia is that it already almost exists— *in reverse.* We have so much energy that we can afford to burn it going *nowhere.* While the United States embraced the idea of importing oil, we've been ambivalent about limiting how much we use. The result is that I'm sitting in Halbouty's office literally shivering from the cold. Little green phytoplankton souls are screaming.

"People. Don't. Care," he says, whacking his chair between words. "As long as they can pull up to the pump and say 'Fill 'er up,' they don't care! Some of the senators from up East are just satisfied with imports. This country has never had an energy policy!"

Halbouty died in November 2004. I have thought of him often since we spoke. Mostly I remember him saying "People. Don't. Care." He's right, of course. Nobody frets about energy when it's cheap. We're still living in the world of the Black Giant, an era when controlling excessive oil production is the main problem. We haven't been forced to come up with a new paradigm. The big question of our time is whether we will change and finally leave the Black Giant behind willingly, or whether we'll just say "the hell with it" until something happens.

Part of Halbouty's argument is that the United States is better off producing energy domestically than importing it. If you're talking strictly about price, he's wrong: Oil from Saudi Arabia and elsewhere is far cheaper than the stuff they're coaxing out of the ground in Texas. In the Persian Gulf, the lifting price (or cost of pumping) can be as low as $3 a barrel, but in the Gulf of Mexico, deep water and deeper oil make for $35-per-barrel lifting costs for some wells. One older study claims the investment cost for finding and getting a barrel of oil out of the ground in the United States is sixty-nine times the cost in Saudi Arabia. But price is a notoriously bad measure for the value of oil—as any Texan can tell you—because it fluctuates wildly from too cheap to too expensive. The lesson of the Black Giant that's been easiest to forget is that government intervention was required to stop careless drilling, waste, and mass bankruptcy in the Kilgore field.

Halbouty's point, too, is that insecurity is a hidden cost of imported oil: It's another way consumers are always paying less for oil than we should be, whether that's paying less money or less attention. Insecurity is tough to quantify. But it just so happens that one of the few manifestations of U.S. official energy policy, the Strategic Petroleum Reserve, is located just an hour or so away on the Gulf Coast of Texas. Along that humid coast, I hope to get a tangible sense of the risk Halbouty is talking about.

## Rig: Last night

The drill bit is at 11,042 feet. The Fifth Avenue has moved.

C. D. looks healthy. He leaves the trailer holding his car keys and sits in the Fifth Avenue with one foot out the door. The engine turns over, whirs, and kicks in. The car stops being a wreck and returns to its rightful place as a comfort machine, with billowy leather pillows attached to the seats.

C. D. has given me his love of hydrocarbons for their own sake—not for the money or the macho, but for the extraordinary chances of their formation and the ingenuity required to get them out of the ground. I've realized that I *was* prejudiced against the extractive industries.

Before I leave the rig, there's one more thing C. D. wants me to see: "At midnight there's juice."

At midnight we climb to the doghouse. The crew has changed and the Worm, the Midget, and Lost and Found have been replaced by new guys. Joe, a family man in his forties, is methodically cutting up carrots under the 1-800-BLOWOUT sticker. Next to him is a twenty-pound bag of carrots, some celery, parsley, oranges, apples, and cilantro, as well as a juicing machine. He sees me and suggests that I buy my own Juiceman Jr. from Wal-Mart when I get home. Who needs Caliprunia when you've got juicing in East Texas?

The new driller is a forbidding man. In a bar, I would give him a wide berth: The bulging muscles on his bare arms sport aggressive blond hairs. Again, first impressions are wrong. "I lost fifty pounds from the juice," he says, inclining his head shyly, "and boiled fish. Makes you feel good." The guys stand around talking about juice—celery versus apple. Parsley versus radish. Joe says he's been juicing for two years, "It's like I can't go without it."

Joe hands me a foamy cup of carrot-parsley juice. I stand on the rig floor with my back to the turning drill pipe and look down at the ground, at the trucks and cars facing outward, ready to escape. One of the new rig hands moves next to me. "I spent thirteen Christmases in one of those trailers," he says without resentment. "My dad was a company man."

**The drill bit** found gas at 13,180 feet. By the standards of the current boom, it was a very good well: a year's production could meet the annual natural gas needs of 13,443 American households. (A single day's output was around three million cubic feet.) Within a few months, the derrick and the crew had moved on across the fields to the next well. C. D. moved out of the mud logging hut to go to a big oil rig in West Texas, where he works as a drilling consultant.

# 5  STRATEGIC PETROLEUM RESERVE

The United States stores about two months' supply of oil in the Strategic Petroleum Reserve (SPR)—700 million barrels in four places along the Gulf Coast. If anything would give a person a sense of how large our needs are, and how deep our fears, it would be the SPR, but unfortunately it is several thousand feet belowground, folded within salt caverns and far out of sight. Recently even the part that lies aboveground has become invisible as the locations have disappeared from maps, Web sites have been stripped of interesting pictures, and the whole thing has retreated under a salty cloak of secrecy because of fears that the reserve itself could be a target for terrorists.

How to get in there? I send some e-mails and make some phone calls and finally speak to a woman in Louisiana who says the sites are not open to journalists. But, she says after a minute, if they were open, the only way in would be through Drew Malcolm with the Department of Energy in Washington.

I call Drew but he is not encouraging. "There's nothing but pipes coming out of the ground and going back in," he says with finality. He mentions that he's suffering from a bad flu and a botched root canal, and he says there's practically no chance I can get permission to enter. I ask if I could go stand on a hill nearby. That would be unwise, he says. If I get within sight of the reserve, I will be greeted by two German shepherds who are "very nice when the muzzle's on." If I persist in looking at the site, say, inside my rental car, with the dogs launching themselves at the windows and biting off the mirrors, armed guards will get in their Humvee and drive out for a little look-see. He doesn't need to mention that the guards have crew cuts and suspicious eyes. "They really like that Humvee," he adds. "They freak out and I don't blame them. They're sitting on top of hundreds of millions of barrels of oil."

In a terror war that exists as much in imagined catastrophes as real ones, the SPR is serving double duty as a defense and a target. After the 9/11 terrorist attacks in 2001, President Bush ordered the Department of Energy to fill the reserve with an extra 100,000 barrels of oil a day just in case terrorists attacked the nation's oil supply. The SPR began to swell, but the bigger and more important it became, the more invisible and inaccessible it had to be.

Despite its attempt to disappear, the SPR has also managed to become controversial to the small crowd of politicos and energy analysts who care about it. Originally built in reaction to the Arab oil embargo of 1973 and as a shield against OPEC's power, the SPR was supposed to be an emergency stockpile—like those 5-gallon jars of peanut butter survivalists keep in their closets—intended to insulate the United States from the politics of keeping foreign oil flowing, to make us self-reliant, and to cause any rogue to think twice about shutting off the nation's oil supply. But thirty years later, the SPR itself has become a political football, a security issue, and a possible hazard in the oil markets. Some economists even say its existence is inflating oil prices by 25 percent. Is it possible that simply by filling the reserve the United States was transmitting fear to the market instead of confidence? Somewhere along the way the SPR has become a 700-million-barrel enigma.

Beyond pipes, I can't imagine what I'll see inside the SPR, but I still want to get in. The little I'd been able to find on the salt caverns that hold the oil suggests that they are the ninth wonder of the world. The only reason I think I have a chance is that Drew seems to take some pleasure describing my dismemberment.

"I like pipes," I say lamely, hoping to wear him down. Drew changes tack. He mentions that he's on painkillers for the root canal, then begins to talk about the birds of the SPR. There's an endangered bird that's nesting above one of the salt caverns, preventing a renovation. "Some bird no one gives a flip about," he says, "holding up the whole thing!" This seems to trigger a free association about another bird. Once, he says, he was in the SPR when a "small attack bird" chased a sparrow so viciously that the sparrow slammed into a car windshield, killing itself. The attack bird, the staff said, was not only extraordinarily ruthless, but when it was through with the killing, it hung its victims on the perimeter fence. What kind of prey? "Sparrows, frogs, snakes, large bugs . . . far out." So far out that it's hard not to attribute some kind of sinister human inten-

tion to the bird, which is exactly what Drew does, shouting "They DIS-
PLAY their KILL!"

The bird seems like a freakish fit for the Humvee, the dogs, and the
sheer volume of oily paranoia in the SPR. The only thing I want more than
to visit the SPR is to visit the SPR with a fenceful of crucified frogs casting
long froggy shadows across the caverns. My heart is beating very fast, perhaps as fast as the heart of a sparrow being chased into a windshield.

"I'd really like to see that bird," I say wistfully. Drew says he'll see
what he can do.


**I dig around** on the Internet, looking for old pictures of the SPR sites.
The Gulf Coast is the most important oil hub in the United States, processing about half the oil and gasoline the country uses. A snarl of
pipelines crawls out of Texas and Louisiana toward New York on the east
and Salt Lake City on the west, as well as everywhere in between. The
four sites of the SPR appear as tiny dots on the Gulf Coast, unremarkable
except in times of emergency, when their oil would be put in the
pipelines and tankers toward refineries around the country. By tanker,
New York is six days from the Gulf Coast; California is sixteen.

Aerial photographs of the SPR sites are bland. Gray squares of concrete sit beside round tanks. The sites appear inert, like dusty piles of
children's blocks. If you scour the photo, you might be able to find a croc-
odilian oil barge lurking offshore where the terminals for loading and unloading tankers are.

I have a hard time imagining the salt caverns full of oil, but one day
I happen across sonar images of them, done by the Sandia National Laboratory.

The caverns appear as maps of connected points. If you look at them
from the right angle, they pop into 3-D perspective, revealing that the
caverns are not really cylinders but long lumpy things of irregular circum-
ference. An image of the Bryan Mound site looks like a bunch of bulging
tube socks dangling from strings. The socks are the caverns and the
strings are the pipes that lead to the surface.

While I wait for Drew to call, I look at those sonar images, trying
to figure out what they remind me of. One day it's "collapsed lung."
Another day: "carrot," "bedpost finial," "finger wart," "Popsicle." It all
becomes a kind of Rorschach game, but the images have a spooky ap-

peal, and finally they most resemble a monument or a memorial to something.

**In a sense** the SPR is a monument to the cataclysmic oil crisis of 1973. Like all monuments, it documents both the size of the shock and the sincerity of emotion—700 million barrels! It's also become a bit of nostalgia, out of step with the world around it.

The oil crisis had its roots in the Black Giant, the big Texas field. Washington policy makers watched it deplete rapidly during the 1930s. As World War II ended, they imagined the United States with a little hourglass full of oil, rapidly running out. Estimating that the country had enough oil to go it alone for just two years in a protracted war with the Soviet Union, they began to pursue a deliberate strategy of importing oil from Middle Eastern countries to make our oil supplies last longer and to flood Middle Eastern kingdoms with cash to keep the Soviets at bay. Some postwar theorists even supported importing Persian Gulf oil because it created a sort of proto-SPR by leaving American oil in the ground as an emergency stockpile. The government lowered taxes on imported oil, making U.S.-based independent oil operators (like Michel Halbouty) furious. Importing oil from the Middle East, with tax breaks, was much cheaper than drilling here at home. Between 1960 and 1973, oil production in the Persian Gulf tripled. Without paying much attention, the United States became increasingly dependent on foreign oil, and by 1973 imports made up a quarter of oil consumption.

During the 1960s Arab oil producers began talking about how to step out of the role of willing and seemingly weak suppliers of oil in order to influence U.S. policy toward Israel. Since 1967 the Arab countries had been discussing whether they could show their power, and change U.S. and British foreign policy in the Middle East, by using the "oil weapon." They decided to punish President Nixon for supporting Israel during the Yom Kippur War. On October 17, 1973, they embargoed oil going to the United States and Britain.

In the melee, the United States lost about 7 percent of its supply, but government allocation programs worsened the shortfall, and Americans soon found themselves standing in line to buy gas. The gas lines incubated a national loss of identity with geopolitical implications. The whole definition of being American was that we drove our cars anywhere

we wanted to. Public transit and waiting in line was something commu-
nists did. In gas lines, people turned their anger on each other, on the
multinational oil companies, and on an imaginary villain named the "oil
sheikh," who was usually characterized with an oil can in one hand and a
wad of cash in the other, grinning maniacally. The British were horrified
to hear an American official speak casually of invading Saudi Arabia.
Nixon's government, addled by Watergate, dithered for the five months
of the embargo, and the president made a depressing spectacle of not
lighting the White House Christmas tree.

With the oil crisis, the long upward expansion of the U.S. economy
since World War II ended and started to reverse. Over the next four years,
11 percent of workers in manufacturing positions lost their jobs. The blue-
collar economy reorganized, and many workers went back to work in
lower-paying positions. There was hardly a family that wasn't profoundly
affected by the crisis. People stopped driving, and when they could they
bought smaller cars. The fact that the oil shortage brought about a deep
psychological insecurity was not missed by the comptroller general, who
mentioned "an illusion of U.S. impotence" in a 1978 report on the crisis.

A giant emergency stockpile of oil seemed to be the answer to oil insecurity. A hoard of petroleum couldn't bring back cheap gas or big cars,
but it could insulate the economy from a shock. Congress viewed it as an
oil weapon to combat the Arab oil weapon: It seemed to offer a cure for
impotence (or at least the illusion of a cure). Senate hearings to plan the
reserve said its purpose was to "provide credible evidence that the U.S.
has the will to insulate its energy economy from major supply disrup-
tions." The SPR would be about "will" as much as it would be about oil
supplies. Like a gunfighter in a movie, we hoped to walk into the bar,
flash a six-gun, and watch bad guys crumble.

And it would be a very big gun—a billion barrels, enough at that
time to buy ninety days of diplomacy, war, or whatever it would take to
get the oil running again. The SPR was cold war thinking translated to
oil: To this day conservatives from the Heritage Foundation compare the
deterrent power of the SPR to a nuclear weapons stockpile.

The building of the SPR, meanwhile, offered opportunities to the opportunistic. A union of tanker ship crews suggested, self-servingly, that
oil be stored in old tanker flotillas—tended by tanker crews, of course.
Another early suggestion in Senate hearings was to put it in large rubber

bags. (Picture hot water bottles the size of airplane hangers scattered across the landscape.) When the committee finally chose to store the oil in underground salt caverns in the Gulf Coast, the governor of Louisiana grudged, "If the federal government is going to pour money down a rat hole, I'd just as soon it be a rat hole in Louisiana." Over the next twenty years, the United States dumped more than $37 billion into the creation, filling, and maintenance of the rat holes.

The second oil shock hit in 1979, before the SPR was finished. The Iranian Revolution and the taking of U.S. hostages cut the world's oil supply by only 4 or 5 percent, but panic sent prices up 150 percent. Jimmy Carter, writing years later, said that Americans "deeply resented that the greatest nation on earth was being jerked around by a few desert states." He understood that people wanted dramatic action, and he called energy security "the moral equivalent of war."

Behind the rhetoric, though, a more effective oil weapon had taken shape. Conservation laws enacted by Presidents Ford and Carter reduced Americans' per capita oil consumption by 23 percent between 1978 and 1983. Coal, natural gas, and nuclear energy replaced oil for making electricity. Cars became 40 percent more efficient, and everything from refrigerators to dryers followed. By 1986 the economy was making every dollar of gross domestic product (GDP) with nearly a third less energy.

And by the time the SPR was ready in 1984, embargoes had become more or less extinct because oil-producing countries began to trade their oil on open markets in London and New York, radically changing the flow and control of oil around the world. Now oil was sold to the highest bidder. Because oil consumers were using less, there was oil to spare. Meanwhile, a concerted effort on the part of the United States to bring oil from non-OPEC nations into the market was bearing fruit. Oil producers could no longer dictate who got their oil and who didn't.

"In a way the SPR is an anachronism," says Sarah Emerson, an analyst at Energy Security Analysis, Inc. "When it was set up, the price of oil was controlled. What mattered wasn't the price; it was volume in the right place. But the '80s threw the market wide open." Today, in a crisis, oil is still available, but it is expensive.

So with the threat of an embargo gone, what is the SPR for? One obvious option would be to use it to keep oil prices low, by putting oil on the market whenever prices crossed a set threshold. But twenty years later, there is no policy for that. "The SPR is smack in the middle of an ongoing debate about whether the government intervenes in markets or leaves them alone," says Emerson. "But the SPR turns market forces on their heads."

Emerson calls the SPR a "highly politicized tool," noting that when President Clinton released oil from the reserve during a heating oil shortage in 2000, he was accused of pandering to voters. The only time the reserve has been used for a politically related crisis was at the beginning of the first Gulf War in 1990, when oil was released well after prices had spiked. Without a clear policy, there's a taboo against releasing oil from the SPR.

Interestingly, the SPR is more likely to be used for weather supply issues than for more political ones. After Hurricane Katrina, the energy department lent 12 million barrels of crude oil to refiners, with the understanding that they'd return the 12 million barrels with interest when supplies were better. But in 2002 and 2003, when political problems in Venezuela reduced the country's exports to almost nothing, President Bush didn't touch the SPR.

The taboo against using the SPR may make it ineffective even at doing what it's supposed to do—riding to the rescue with millions of barrels of oil when we need it most. A 2005 simulation called the Oil Shockwave had a group of former cabinet officials try to contend with a 4 percent fall in the U.S. oil supply. The players concluded that the SPR's policy was too convoluted to create a consensus on how to use it. In the confusion, the (simulated) price of oil went to $160 a barrel.

Ultimately, the domestic politics of the SPR are so tricky that it tempts leaders to turn to oil politics—the very thing the SPR was created to prevent. "Instead of using the oil in the SPR," says Emerson, "the Bush administration asked Saudi Arabia—with whom we have an increasingly deteriorating relationship—to supply a million barrels a day more oil before the beginning of the Second Gulf War. That ought to give us pause."

What happens when you have 700 million barrels of oil and no plan to let go of it?

**Drew calls** to say I have permission to enter the SPR, but only for an afternoon. He tells me to meet him in the lobby at the Comfort Inn in Lake Jackson, Texas. "The attack bird," I ask. "Will I be able to see it?" Yes, he says, the bird will be there.

Lake Jackson is southeast of Houston. It's a sullen city surrounded by refineries and scrubby bird sanctuaries—each forbidding in a different way. The town was designed as a bedroom for the workers at the petrochemical plants of nearby Freeport, but its founders didn't have grand ideals. I pass a street named "This Way" and another named "Pin Money." Small houses on blocks are dwarfed by the Ford 250 pickup trucks parked next to them. A sign in front of a church, in the place where you'd expect to see a Bible verse, says STOP GLOBAL WHINING in white plastic letters.

Drew turns out to be very tall, with amiably hunched shoulders and a slightly weary look. He is apparently off the painkillers, and self-assured enough to arrive in Texas wearing a pastel jersey, loafers, and patterned socks. He confesses that he wore loafers and patterned socks to his Department of Energy job interview. "I was in a mood," he says.

When he went to work there, Drew says, he didn't know what he was getting into—he'd retired from navy public relations and gone to work for Grumman. "I was new meat," he says, "so they made me the SPR boy." Apparently, being the SPR boy was considered tedious, but Drew quickly discovered that the job was mysteriously powerful. One of the first things he learned was that he should make all announcements about the SPR after the oil market in New York was closed because some information could create chaos in the market. "The first time I heard that I thought Wow, I never thought of that before," he says.

What Drew didn't realize, and is probably too modest to admit, is that by becoming the SPR boy, he became an extraordinarily important figure in the petroleum markets, and thus an important person in your life and mine. In controlling information about the SPR, he controls information about a reservoir of oil that is equivalent to three times the stocks of oil held by refineries and industry in the United States. The lack of a clear policy (or prices that "trigger" releases of oil) means that information about the SPR is watched jealously. When the reserve released a comparatively puny 17 million barrels of oil in January 1991, for the first Gulf War, prices fell from $27 a barrel to $21.

If Drew were really to take a lot of painkillers and announce that the United States was releasing a few million barrels of oil in a short period of time, he could cause a disaster somewhat worse than an oil shortage. When oil prices are very high and refineries have both borrowed money to buy oil and hedged their investments with futures purchases, a poorly thought-out release from the SPR could leave U.S. refineries $200 billion in debt, according to energy economist Phil Verleger. A significant bankruptcy of refineries could cause more chaos than a temporary shortage of oil. And here sits Drew with his loafers splayed out like duck feet.

To be fair, Drew insists that the SPR is about "emergency supply," not "market manipulation." That may be true, on the face of it. But the mere fact that the SPR exists makes it a market player, and Drew's job is all about being the lubricant between the twin big wheels of government and the market.

Drew says that when he first got his job, he had to "run around the building, standing on desks and saying pay attention to me." At two in the afternoon, he'd sit panic-stricken at the computer writing a statement for the energy secretary to read on Fox that night. He learned to put information out in the least sensational ways. When oil was "swapped out" of the reserve—essentially lent to oil companies, which would then repay with extra oil as "interest"—he had to present the information in a way that would not shock the market or cause gossip. "You have to be very careful of giving false impressions," he says.

Drew also has to clamp down on political appointees to the Department of Energy who get the urge to make some sort of energy-related pronouncement. "They're like two-year-olds," he says, "saying 'MINE.'" He dissuades them from speaking. "It's much too important. It's not something you should play with."

How does he avoid getting crushed by the appointees? "I'm pushy," he explains, putting his feet on the coffee table, "and I fight and get shot down. I figure you give it your best shot and salute and shut up. I'm an old military guy."

In late 2002 the price of oil began to rise dramatically after the Venezuela crisis. An investigation by the Minority Senate Investigations Committee laid the blame on the Bush administration for continuing to fill the SPR. Bush's bluster, they said, was costing U.S. taxpayers an extra $1 billion a month and inflating oil prices by as much as 25 percent.

The report, however, didn't get much attention, probably because its 400-odd pages of plot and counterplot induced migraines.

When I asked Drew about the report, he first said he hadn't heard of it. Then he said, "We haven't spent money on oil since the early '90s." He says the SPR is being filled with oil owed to the government by companies that produce oil from federal lands. This is true, but what he doesn't mention is that the oil taken out of the market does affect its behavior. A decent distance between government actions and market forces is a fiction Drew is charged with maintaining, and he does a good job of it.

Most of the reporters who call him are not interested in subtleties but in broad conspiracies. "From my chair in Washington it's astounding the rumors that get started," he says. The worst rumors, he says, come from South America, which has a long history of distrusting the power of the United States. "It's America, God knows. We've got money and a vast reservoir to fill and discharge at whim," he says sarcastically.

Drew thinks the SPR attracts conspiracy theories because people don't understand what it is and how it's used. "The SPR is like Fort Knox," he says. "Years ago, when the economy was all about gold, there were always rumors about Fort Knox. 'A friend of a friend of a friend told me . . .' The SPR has taken on the same mystique, and it's increased because you can't see or touch the oil."

And the bird? Well, it seems Drew's been thinking strategically all along. Far from being sloppy on painkillers, and talking idly of birds, he's angling to get more press coverage of the SPR, so that it doesn't appear in the news only when someone's got "an agenda."

**Bryan Mound** is outside of Lake Jackson, where pale landscape hits pale sea. The entrance is a parking lot and a faded checkpoint. The air is so thick and salty it almost scratches. Nothing suggests that more than 200 million barrels of oil—with all their economic, social, and military implications—are underfoot.

The mound's elevation is just eighteen feet; anywhere else it would be just a hump. Through the checkpoint's gate I can see what looks like the sports field of a poor high school: dull green grass interrupted by bald patches. Pipes extend into the air from the bald patches and end unimpressively with a few bolts or valves. If the pipes were athletic equipment, I wouldn't want to watch the sport. There are no frog carcasses hanging on the fence. Not even a grasshopper.

"Oil?" says one of the guards in the entrance. "It smells bad. And it comes out of the ground. Hnff. What more do you need to know?" My bag is X rayed, a wand run over my shoes. A digital scan of my driver's license appears on the computer screen. The guards are carrying rifles, but the general tenor of things in the scan shack is sun-baked and uncurious. Forget conspiracy theories and Senate inquiries: The SPR is nothing more than a place to store a lot of oil. And it smells bad.

"I know you'll love this place before you leave," says the manager, a man in his fifties who seems somewhat embarrassed to be proud of this scraggly mound. He leads us across an enormous carpet displaying the seal of the Department of Energy and into a low building.

Resembling an elementary school from the 1960s, the building harbors that same sort of comfortable bureaucracy—stuffed animals perch on the computers. I've been told I'll see a video, and I'm expecting a dark closet with a TV in it, but the manager brings me to a room where fourteen uniformed reserve workers sit around a table, examining me with interest. I sit in the empty chair in front of a hard hat labeled GUEST, and try to look unruffled. They offer me pound cake and a soft drink. Later I wonder if pound cake is specified in one of their procedures for receiving guests.

As it turns out, the SPR is the last place to contemplate whatever chaos or Armageddon would result if the United States had a full-on energy crisis caused by war or catastrophe. For one thing, they never use the word "crisis" here. Instead they say "supply interruption."

And the very idea of the so-called interruption has been so securely encapsulated in procedures, flowcharts, and schedules that it has become incidental to the greater business of following procedures. There is a procedure for checking the pressure in the caverns and a procedure for printing out the pressure checks, then another procedure for handing the printouts to the librarian at the end of the day. There is a procedure for removing raccoons from the maintenance shed. There are procedures for developing procedures for removing mockingbirds from the engines of the oil-spill cleanup boats in the event one needs to follow oil-spill procedures. In short, the SPR is the Boy Scouts gone mad. They are always preparing to be prepared.

There is a script for the manager's PowerPoint presentation, and he doesn't want to leave it. "The mission of the Bryan Mound SPR Site is to develop and maintain a 232-million-barrel crude oil reserve and provide a drawdown capability of 1.5 million barrels per day within thirteen days of notification," he reads.

If a supply interruption did occur, the president would have to decide whether it warranted releasing oil from the reserve. Once he made the order, an elaborate flowchart mapping out days 1 through 110 would kick into action. Within the flowchart are some interesting caveats regarding energy security and the SPR. For one thing, there aren't "57 days" worth of oil. There is a month's worth of oil, pumped out at full speed, but after that the volume of oil the caverns can pump out drops off. A full drawdown would actually take more than a hundred days of dwindling flows. In addition, the SPR has never really been tested—previous drawdowns have been small. So it's not clear how well it would actually work. Would there even be room in the pipelines near the reserve sites? Finally, the caverns in the SPR contain different kinds of oil—sour (containing sulfur), sweet, and different grades. So the reserve needs to mix the oil before shipping it off to refineries that are equipped to deal with the available grades and degrees of sour. In a dire situation, would everything work? The people here devote their days to thinking it through in advance.

Life in the SPR, though, is made up of little familiar pleasures rather than disasters. Around the table sit the bird and animal man, the security man, the cabin pressure guy, the pipeline integrity guy. . . . When the slide showing a bobcat swimming into the reserve appears on the screen, everyone giggles with modest pride. These guys have to be expressly forbidden to feed their lunches to the alligators. When Mike, a wiry man with a Fu Manchu mustache and iridescent bubble-shaped safety glasses, says he's in charge of "all the wildlife that's not employees," he seems to mean it. He later speculates on whether the site is haunted.

We leave the office to look at the meter skid, which measures the oil entering and leaving the reserve. It's a Seussian grid of red pipes with bolts the size of cupcakes. A three-foot-wide pipe is labeled CRUDE. This pipe is as close as I'll get to understanding the volume of oil here. The guys say the pipes don't make any noise when they're filling. The whole process is silent. "Doves love the meter skid," adds Mike.

The best parts of the SPR lie deep in the salt below us, where the oil sits in hollowed-out cavities. When the sea evaporated during the Jurassic era, it left behind a layer of salt that may have been 5,000 and 6,000 feet thick. That salt is known as a "mothersalt." (This mothersalt is named Louann. I picture her holding a fresh-baked pie.) Eventually Louann was covered with layers of rigid sediments. Among them, Louann began to insistently push her plastic self sideways and upward. Faults happened, folds happened, depositions happened, and Louann's long salty fingers hit the sediments under the Gulf Coast, causing little domes like Bryan Mound to appear.

The caverns themselves are "mined," by using water to dissolve and remove the salt. Once hollowed out, the caverns, like tanks, can be filled with oil or chemicals, and they have the advantage of sitting thousands of feet below ground level. The four SPR sites have sixty caverns, each holding a bit more than 10 million barrels of crude. The average cavern is 2,000 feet long and 200 feet in diameter, which makes them taller than the Empire State Building. Imagine the caverns as skyscrapers hung up-side down underground, and you have some idea of what they're like. The salt walls of the caverns probably even have some of those buildings' deco grandeur: In core samples, the salt crystals glitter icily, interspersed with small shiny bits of radioactive shale.

When we talk about the caverns, it's easiest to compare them to tanks, but that's not accurate. Salt is an extraordinary container. It's soft and malleable, like children's modeling clay. Put oil inside the mushy salt, and the salt will embrace the oil. If the oil pushes against the salt, the salt will not let it escape. Geologists call salt's behavior "self-healing." Salt's agreeability means that the caverns don't leak and don't need maintenance, so storing a barrel of oil at Bryan Mound for a year costs only 20.5 cents. (According to the SPR workers, Japan spends $10 per barrel per year to store oil in rock caverns and metal tanks.)

The process of removing the oil from the caverns is equally elegant: Oil floats on water, so pumping water into the base of the cavern will make the oil rise to the top. The caverns also possess a marvelous built-in stirring mechanism to keep the oil circulating. The bottoms of the caverns are 123 degrees F while the tops are around 120, thanks to the geothermal temperature gradient. The oil in the caverns stirs itself by convection, as the warm oil rises to the top in a languorous cycle.

In the Department of Energy's SUV, we go on a tour of the grounds, past a heat exchanger, which looks like a giant radiator, and past more bald patches and more Christmas trees. Pipes, where I can see them, are pale yellow, gray, or red. There's an 8-million-gallon tank of water where a confused catfish once ended up when it swam through the water outlet pipe. It had to be fished out.

I think I can now confess that the manager is right, I do love the SPR. It's very similar to heaven in that Talking Heads song: a place where nothing ever happens.

Except, of course, that something could happen. Something big. And for that reason, one third of the mound's 120 workers are security.

**Four security** guys sit around the table, but only one will talk. His name is Buddy, and he looks like a younger C. Everett Koop, right down to the beard and the white military shirt. He is willing to acknowledge that the German shepherds were trained in eastern Europe. He also allows that no one has ever entered the site "on purpose." Whether that means they've entered while drunk or they haven't come in at all, Buddy refuses to confirm or deny. I want to ask if his job isn't awfully existential—all this waiting around for no one to show up—but before sliding out of the room Buddy says he feels like he's "got a screw on his thumb" and he's "had a heart attack already."

The guys left at the table say they're not too worried about people entering the SPR aboveground. What they really care about is someone drilling into the caverns from below. A knowledgeable person could set up a drilling rig nearby and drill horizontally into a cavern to get millions of barrels of oil for free. This sounds like a fantastic plot for a caper movie: Three good-looking guys, a drilling rig, and and a plot that hinges on some quirk in the geothermal gradient. But apparently the SPR's obscurity has prevented us from seeing George Clooney in the role of a horizontal driller.

To prevent heists, the reserve does frequent "flyovers" to monitor who's drilling where. Four times a day the pressure in the caverns is checked to see if a drill bit has poked through. The control room, which has one wall of glass windows, resembles old NASA: Lights blink and volumes of data spew from a printer into a box on the floor. Aboveground security, though, seems to be less of an issue. When mockingbirds nested in the junction boxes, the environmental guys bartered with security to be reasonable. They let them stay for eight weeks.

Drew gives me the signal that it's time to leave. The bird, I say. You have something that displays its kill on the fence? Oh, yes, says Mike, Fu Manchu bobbing enthusiastically, the shrike. "Let me tell you, if that thing weighed a pound, it would be a WMD," says someone. Mike says the shrike doesn't have talons to rip apart its prey, so it hangs them on the fence to "tenderize" them. What gets hung on the fence? "Mice, grasshoppers, small security guards."

**I drive** away from the SPR, past the Gulf Coast's bristle of refineries, trying to sort out what kind of security the reserve actually offers. The most obvious problem with the SPR is that all the sites are on the Gulf Coast—which happens to be the most vulnerable place in the country for hurricane damage, and the place where more than half of our oil originates. In the case of a weather emergency, both U.S. supply and backup are in the same place.

A second problem with the reserve is that the pipelines leading out of it make good targets for terrorists. In 1997 police arrested members of the KKK who intended to blow up a gas pipeline to finance other terrorist activities. In 1999 Vancouver police arrested a man who had explosives and enough timers for fourteen bombs. He was planning to blow up the Alaska pipeline so he'd make money in oil futures. The pipelines leading out of the Gulf are particularly vulnerable. In the early 1980s a report on U.S. energy security noted that one person could destroy three of the Gulf Coast's pipelines in three nights. By 2002 pipeline companies realized they were vulnerable to computer hacking and electrical interruptions. The federal agencies that regulate the pipeline industry are understaffed and have left pipeline safety largely at the discretion of pipeline operators for years. As of 2004, the Department of Homeland Security devoted just five staffers to pipelines.

Even if everything goes well in the Gulf Coast and the pipelines, there is still another problem. Together, all the SPR sites are capable of pumping out only 4.4 million barrels a day when they're running at full tilt. That's about a third of the quantity of oil that the U.S. imports daily (assuming that there's even that much room in the pipelines to carry it). Adding potential oil imports from Canada and Mexico to the SPR, you still end up with a 4.3 million barrel shortfall—nearly 20 percent of the oil the United States consumes daily. That's three times the size of the 7 percent cut that triggered a crisis and economic restructuring in the 1970s. Even if the SPR works perfectly, it still doesn't protect the economy.

The 2005 Energy Bill calls for expanding the SPR to 1 billion barrels of oil—but it will be in the same sites in the Gulf Coast, and connected to the same pipelines. Instead of 700 million barrels of oil without a plan, it will be 1 billion barrels of oil without a plan.

The SPR offers the illusion of safety but no real insurance. It is too small, too centralized on the Gulf Coast, and too vulnerable to be effective. Insulating the United States against oil shocks requires more work than just stockpiling. Just as the conservation measures of the 1970s weakened oil producers' power, the country could begin reducing oil demand to insulate the economy. "For the past two decades everyone thought problems with oil were a figment of someone's imagination," says Oak Ridge National Laboratories energy economist David L. Greene, who makes a strong case that both history and economic theory indicate that reducing demand makes for lower prices and smaller shocks, and decreases the power of cartels.

How would we go about cutting oil use? It might not be that difficult. In the summer of 2005, the International Energy Agency did a series of forecasts on saving gasoline in a hurry. Reducing the highway speed limit to around 50 mph, it found, could save 738,000 barrels a day. Building an infrastructure for carpooling could save another 770,000 a day. Encouraging telecommuting would save another 538,000 barrels a day. Getting people to inflate their tires correctly would save another 154,000. In other words, we could save 2 million barrels of fuel a day—about half of what we could expect from the SPR running at full blast—and we'd save money. Enacting laws to increase cars' fuel efficiency by a small amount could save another 2.5 million barrels a day by 2015, which is probably faster than anyone could get the new SPR up and running. Small plants producing biofuels or synthetic fuels from coal could also step into the breach in a crisis. Their advantage is that they could be scattered around the country outside of the Gulf Coast, and they'd increase alternative fuel stocks in general. In short, there are dozens of different strategies and ways to spend $20 or $30 billion besides pouring it down the rat holes of the SPR.

**In the aftermath** of my visit to the SPR, I regret that I never got to see the shrike, so I do some research. In photographs, the loggerhead shrike is a robin-size songbird wearing a Lone Ranger mask: cute, round, and dusky, with a hooked beak.

But appearances deceive.

The shrike's other name is the butcherbird, for its methodical way of ripping the eyes and mouth of its prey before moving to the neck and the guts—after they've hung on the fence for a while, of course.

The shrike is always misunderstood.[1] It is not a villain or a psycho-killer, but a collector. Shrikes have been known to hang a frog on a fence and return to snack on its mummified corpse eight months later. They like to kill and store more things than they can eat. Scientists have discovered shrike "larders" containing snakes, fish, mice, many species of birds, and insects along with frogs, lizards, bats, snapping turtles, shrews, voles, and a hispid pocket mouse.

Think of the shrike as a conspicuous stockpiler, an obsessive-compulsive hoarder with an exhibitionist bent. It seemed fitting that the shrike lives in the SPR. Like the SPR, the shrike keeps a stash, but flaunts it, even when it has no intention of using it. The question is: What effect does an ostentatious stockpile have on the world around it?

**Between November** 2001 and mid-2004, crude oil prices rose from around $16 a barrel to more than $40. It was a long, slow, incremental rise. During those years the United States placed more than 90 million barrels of oil in the SPR.

In the markets, a growing number of people believed that the SPR was having a mysterious pull on oil prices. In January 2004 Goldman Sachs attributed $2.00 in oil's price to the SPR and another $2.25 to copycat buying from the SPRs of other countries.

"Strategic stupidity," sputters energy economist Phil Verleger, who maintains that holding strategic stocks of oil—or any commodity, including corn and wheat—inevitably pushes prices up. Verleger says that when the SPR was created, its very existence encouraged refineries and

---

[1] Humans take shrikes personally. "The most arrant hypocrite in the whole bird calendar. Its appearance as it sits apparently sunning itself, but in reality keeping sharp lookout for prey, is the perfect counterfeit of innocence," shrieked a shrike critic in the magazine *Birds and Nature* in 1901. "It sings and sings and is a villain still."

oil companies to reduce their own inventories, which not only shifted the cost of carrying the inventories to taxpayers, it also meant that companies bidding for oil had smaller stocks, which pushed prices higher. Worse, he says, the pattern of filling the SPR with oil now rather than buying oil futures (which are discounted for future delivery) means that the treasury is losing at least $8 on every transaction. He says that without economists at the helm, the SPR has pushed up the price of oil in the world market by $8 a barrel.

Other analysts disagreed. I called David Pursell, an analyst at Simmons & Company International, to ask if the SPR fill was driving prices higher. Pursell seemed surprised by the question. "The global demand is 77 million barrels a day," he says, tap, tap, tapping on the keys of his calculator, "so that means the SPR is taking one-tenth of 1 percent. It is inconsequential. If you're looking for confirmation of a conspiracy, I'd say no." The Petroleum Industry Research Foundation, an industry group, says that the SPR fill ads 1 cent or less to the cost of a gallon of gas.

"A hundred thousand barrels a day is less than a drop in a bucket— it's a rounding error," says Fadel Gheit, an analyst for Oppenheimer & Co. Inc. who's been watching the oil markets for thirty years. Gheit has an accent and his voice is well suited to comic exasperation, which is the way he talks about the SPR and the U.S. government. He says the problem with filling the SPR was that it was done in public, which signaled to crude traders and other countries that the United States was afraid.

"After 9/11, part of the Bush strategy on terror was to announce that we were adding 100,000 barrels a day to the reserves. Strategically, this was the stupidest thing. This was just political grandstanding, but it sent exactly the wrong signal and it spooked everyone. China, India, Korea, and the European countries panicked and started buying more oil than they needed—they assumed the U.S. knew something they didn't."

Classic shrikish behavior.

Gheit takes a breath and begins to talk about panic buying. He says that hoarding begat more hoarding, and oil prices climbed higher. "This thing was mishandled royally! Our failed policies and strategies are responsible for $45 oil!"

But for all of his outrage, Gheit has no idea how much of the current price is the result of SPR hoarding. "There's no formula. No science," he says. "With five different analysts you'd have five completely different explanations. It's all about emotion and perception."

Some confirmation that the SPR was perceived to be raising prices arrived in April 2006, when President Bush halted deliveries of just 20,000 barrels a day to the SPR in an attempt to ease crude prices, which by then were $75 a barrel.

**The enigma** of the SPR bugged me. I hated not knowing what it was doing in the oil market. So in spare moments I'd call more analysts, hoping someone would say something definitive.

One afternoon I called Tom Bentz, an oil analyst for BNP Paribas who's been watching the market for twenty-seven years. He's got a warm, relaxed manner, and he doesn't see the SPR as a big player in the market. He runs through various theories about whether the SPR creates high prices or eases them. "I guess it's the perception of each individual that makes a market a market," he says.

It seemed it was time for me to leave the SPR and head for the market to get a sense of all those different perceptions. Perhaps the SPR was going to remain an enigma to the bitter end. I looked forward to the fresh breezes of the New York Mercantile Exchange, the free flow of ideas and commerce, a nice dose of market logic.

Bentz continued in a philosophical vein. "The thing is, you don't see a lot of the SPR. Is it really even there? Who knows?"

# 6  NYMEX OIL MARKET

On the day I dropped in on Tom Bentz, crude oil futures were selling for $45.04, which was the most expensive they'd been in the twenty-one-year history of the New York Mercantile Exchange (NYMEX). But by August 2004 record-breaking oil prices had become a habit, so passing the psychological barrier of $45 was like crashing through a trick plywood wall only to find a giant yellow arrow pointing to the next so-called barrier of $50—something like Mr. Toad's Wild Ride at Disneyland.

When I called Tom to ask him about the Strategic Petroleum Reserve, we digressed into a discussion of the NYMEX crude market. He mentioned that crude prices had become intertwined with his personal life. Name any oil price from the past two decades, he said, "and I can tell you exactly where I was for every high and every low, who was sitting next to me, and what caused the price." He said he'd be willing to make a chart of crude oil prices since day 1 and tell me his story. From there I might make my way downtown to spend some time in the actual NYMEX crude pit, where the price of crude oil futures is set by a few hundred men jumping and screaming in open outcry trading.

Tom is vice president and senior energy analyst with the French-owned bank BNP Paribas. He works on the thirtieth floor of a reddish gray Manhattan high-rise where the brokers spend their days surrounded by computer screens, talking on their headsets to oil companies who want to buy or sell crude oil futures. They send those orders downtown to the company's traders in the crude pit at the NYMEX.

The brokers in his office sport the classic commodity trader's hairdo—longish, swept back from the forehead in smooth, well-gelled ribs, which make them look like otters that have just surfaced from the depths. The upstairs brokers view the action in the trading pit from a physical and psychological remove, remaining cool while the traders on the floor get hot and sweaty. Now that the trading has stopped for the day, the light in the office is purplish and hazy.

At forty-seven, Tom is one of the older analysts, but he shares the hairstyle, the blue polo shirts, and a fresh-scrubbed attentiveness that makes him seem younger. He is tall and restless, often shifting his weight in his chair, leaning forward to make a point, gesticulating and miming as he talks. The oil futures market is for him still an opera, an object of great affection.

He does not seem tired from the day's trading, or even surprised that prices set a record. "It was another all-time high," he says, "but after that the market settled back down." He swings in his chair, opening his hands, as if to show he's not holding any cards. On the table in front of him is his chart of oil prices, a long craggy sawtooth of peaks and valleys starting March 30, 1983, and ending today, August 13, 2004. The chart is marked with bits of blue scratchy pen: "Exxon Valdez 3/89" and "Iraq Invasion of Kuwait 8/90," "Nigerian strike 7/94," "Cold weather record 11/99." And there are personal notes too: "TB [Tom Bentz] cancer 11/01 recovery begins" is written next to a six-month upward trend in oil prices starting in late 2001. He leans back in his chair heavily. "We're just waiting for the next event."


**If the United States** celebrated economic milestones the way we celebrate military and political ones, we'd all get a holiday on March 30 to honor the day in 1983 when crude oil futures first began trading, making all of us participants in the giant world market of petroleum. In the 1970s OPEC supplied 67 percent of the world's oil, and it was able to limit its members' production to keep prices high. Prices then didn't change much because they were set in long-term contracts and no one knew what those prices were, because they were set in back-room deals. With the arrival of the exchange, the old organization of the oil market changed: Information flowed more freely, prices fluctuated, and oil could be bought on the spot market. With new non-OPEC oil producers entering the market, it became much more difficult—and often impossible— for producers to control the price of oil.

You can thank Maine potatoes. The New York Mercantile Exchange opened in 1872 as the Butter and Cheese Exchange of New York. Loads of dairy products, and later eggs, were traded openly, setting prices. After World War II Maine potatoes became, well, the meat and potatoes of the exchange. But by the late 1970s Maine potatoes were getting very unpredictable. The potatoes promised in the exchange and the gnarly tubers that arrived on the train cars from Maine were two different concepts of potato. In 1977 Maine farmers defaulted on the delivery of 50 million pounds of potatoes. An exchange's reputation depends on being able to assure sellers that they'll get paid while promising buyers that they'll get their potatoes. Without the potatoes to back up the futures contracts, the NYMEX was as useful as a pile of moldering spuds.

The chairman called his advisors to his apartment, where they made a list of other substances the exchange had gotten approval to trade over the years: apples, silver coins, nickel, currency, frozen turkeys, and oil. They decided to try heating oil. In the New York area, small heating oil companies bought oil from major companies, and the majors pretty much set the price. The traders thought they might have a good business because with an exchange, the small companies could have more influence on the price and they could buy oil months before they needed it. So, on November 14, 1978, the heating oil exchange opened for business. By the end of the day they'd traded twenty-two contracts, which represented 22,000 barrels of oil. Not much. The problem was that they needed to be sure this oil would be delivered. If heating oil deliveries were as bad as the potato fiasco, they might as well not bother.

Enter Tom, who had grown up in a tough neighborhood in Brooklyn. He worked for a commodities brokerage, hustling up clients for gold on the phone, in his late teens. Then he moved to another firm and got involved in delivering the gold contracts from one commodities house to another. (In the days before electronic transfers, the houses had runners carrying boxes full of stamped contracts to the purchasing commodities houses, where they'd dump the contract through the door.) "After the stuff was traded—it could be gold, coffee, cocoa—you had forty-five minutes to get the contracts to the house," he says, miming frantically stamping the backs of contracts. "It was basically a hot potato."

Tom rose through the ranks, and in 1978 he was asked to get involved with doing deliveries for the heating oil exchange. "When your heating oil contracts started trading, the delivery rules were kind of flimsy," he says. "If you held on to your contracts [the futures] until the day they expired, you were guaranteed delivery, whether you were a major company or some jobber getting 2,000 barrels. I got involved in every aspect of delivery, including literally sending barges to the delivery terminals in the harbor." Sometimes the seller didn't have the oil. Sometimes the buyer wasn't ready. "Many incidents were potential defaults. I gained a bit of a reputation."

The heating oil exchange was popular because small heating oil delivery companies could buy oil without having to accept the prices the majors decided to charge them. They started using the exchange to buy oil for later delivery, and gradually they started to use the futures market to hedge, or protect against price fluctuations. In 1983 the exchange began trading crude futures as well as heating oil, and then anyone in the world could buy crude.

"Before NYMEX, nobody knew what crude prices really were," he says. Because OPEC had sewn up long-term contracts with buyers, the new non-OPEC oil producers like Norway and Great Britain started selling crude on the spot market in the late 1970s. Their prices were lower than OPEC's, so OPEC's customers began to fish around in the spot market—where they could buy a tanker of oil for delivery soon—in search of cheaper oil. By 1982 half of the world's crude was trading on the spot market. Tom moved from the back office to doing trading and analysis. "I fell in love with the energy markets and I knew that was where I wanted to be."

For refineries, buying crude everyday on the spot market is risky, because they can't plan what prices they'll be paying. It would be as if you had to pay a different price for water every day. One morning a shower might cost 25 cents, but the next day it might cost $20, making it difficult to plan your bathing schedule.

On the NYMEX, traders exchange futures contracts—certificates representing 1,000 barrels of oil to be delivered on a specific day between one month and six years in the future. NYMEX contracts are written for only one kind of oil—West Texas Intermediate, known as WTI—and the oil is to be delivered at a single spot—the pipeline in Cushing, Oklahoma.

Buying futures is a way to plan ahead for refiners. Buying oil two months in advance at a known price lets them plan accordingly. (Just as buying next month's showers at $2 apiece would allow you to just get up in the morning and shower, without having to recalculate your budget daily.)

Futures are also useful for oil producers. The problem with owning an oil well is that once the oil is out of the ground, it has to be sold, regardless of price, because storing oil is expensive. So oil producers hedge, insuring themselves against falling prices by selling their oil in advance. In 1985 wildcatter T. Boone Pickens worried that the price of oil was going to fall, so he hedged all of his company's production for 1986 at $28 a barrel. "In the first half of [1986] crude prices went below $10," he writes. "From then on, I was almost continuously in the market." Pickens went on to make billions in energy futures.

Jobbers buy futures to soften the effect of price swings when they're buying gasoline on the spot market. If, for example, they know they have to buy gasoline on the spot market in two months, when prices could be very high, they can buy futures contracts for two months from now, when they're cheaper, and sell them before they expire, when they're closer to the spot price. Then, when they have to buy gasoline from refiners on the spot market, they will have developed a small cushion from hedging in the futures market. Many major refiners also hedge in the market.

And, of course, there are many people trading in the market who hope never to have to touch oil. Hedge fund operators and other speculators may invest in paper barrels and sell them quickly when prices rise. Only 5 percent of the futures contracts on the exchange actually end up being physically delivered as oil.

The power of the NYMEX is that it represents the richest, hungriest oil consumers at this time—the United States. "Our prices will draw oil from anywhere," says Tom. "I view the oil market as a big pie—when you cut a piece the fruit shifts and flows from one place to another. Whatever you take away from one place affects some other place." And so the prices set in New York affect the prices on London's oil exchange and those in Riyadh, in Singapore, and in Argentina. (If prices were much cheaper in London, it would make sense for someone to put that oil on a ship and send it to the United States.) The markets balance out uneven supply and prices.

The futures sold at the NYMEX also act as an index for all the other barrels of crude that get sold in the world. There are 161 kinds of crude oil (OPEC and non-OPEC), and each has a price that is indexed to the cost of WTI sold on the NYMEX. (Brent is the crude used at the exchange in London.) Distances, too, are calculated in relationship to that pipeline in Cushing, Oklahoma, where NYMEX futures claim delivery, even though only a few barrels will ever really be delivered there. As light crude oils become scarce and heavy sulfurous grades become more common, the paper barrels of WTI look even less like the real world of crude.

Before I head off to the NYMEX to see the trading, there's something I can't grasp. The exchange's power is inversely proportionate to its size. While the real oil market requires that 84 million barrels of oil be delivered every day, the number of paper barrels traded at the NYMEX is only a small percentage of the total, equivalent to about 80 percent of U.S. oil production, or around 4 million barrels a day. Those same paper barrels get traded over and over leading to the high trading volume on the exchange—sometimes as much as 200 million barrels a day. Commodity futures markets work by frantically trading an almost symbolic quantity of product, but I can't get over the elemental absurdity of the relationship between the small number of real barrels and the power of the exchange in the real world. I ask Tom to explain it and he describes the exchange as a giant brain. "All of the feelings and information in the world are translated into that price," says Tom. "It's got stories in there about the weather, politics . . . everything. The price is a depiction of all the information that's known."

Tom warns me about visiting the market. "It looks like total chaos. But the job gets done."


**To reach** the NYMEX, I have to pass "the Hole" where the World Trade Center was until September 11, 2001. The pit is still raw on the day I walk around it on the way to One Financial Center, where a trader has agreed to take me on the floor.

The trader has the otter haircut and wears a loose navy coat with orange stripes that looks like something a bellhop might wear in a rural Bangladeshi hotel. We go through the first level of security as we enter the building, and then go through about six more, including three metal scanners and a photo ID badge center, as we ascend to the trading floor. Heading upward through higher levels of security, more men in colored coats join us on the escalators. Red. Green. Yellow like mustard on a hot dog. Turquoise inset with gold mesh. One man's coat is decorated with images of the 1970s cartoon character Fat Albert. Another's is blue with pineapples and an American flag patch. A slot machine with dice. "How's the ring treatin' you?" they greet each other.

The coats distinguish among the dozens of companies and individuals with seats on the exchange. Virtually everyone decorates his coat (traders are mostly, but not all, men) with some kind of 9/11 memorabilia: flag pins; little safety pins carrying red, white, and blue beads; buttons showing either firemen or the faces of people who died. One company has done away with all pretense and just adopted flag-patterned jackets.

The density of people in bright coats increases as we near the door to the trading floor. One of the few women—who's wearing high heels and a coiffure with her bellhop coat—is saying "Yeah. He fucked my balls. What was I supposed to do?"

And then we're on the market floor, and what I see is a dense grid of people crisscrossed by long curly phone cords. Walking among phone cords is like dancing the limbo—step over them, bend under them, do not yank them because attached to each cord is a person curled around the receiver in a secretive fetuslike stance. I don't see the pit until I've cleared the phone cords and crawled up a series of risers among the phone fetuses. From this perch I can see that the pit is in the center of the floor. It's not so much a pit as a set of risers in two concentric horseshoes, like a tiny amphitheater. At the top of the horseshoe there is a round desk with a very thin young man wearing a yarmulke and industrial safety glasses, which jut out an inch or two from the sides of his narrow head, like old-school rapper's sunglasses. In front of him is a computer.

Just before the market opens at ten, the traders jostle into place on the risers, facing inward, their hands clutched at their chests like a gospel choir waiting to wave hosannas. The manager of the Houston Astros opens the exchange, and suddenly all of the men in the ring are yelling and punching the air with hand signals. Their cries start as individual human squawks before they're absorbed into a great roaring rug of squall. Using a private sign language of fingers, they communicate prices and batches across the ring to each other. When they make a deal with someone across the pit, they huck cards confirming the deal into the center of the ring. A squad of men in safety glasses pick up the cards and hand them to the thin man at the computer.

Around the edges of the ring, on another, more substantial set of risers, stand analysts, assistants, and interns, all on phones with long cords, huddled near small stands and cubbyholes for their computer screens. Their cords are long enough so that the people in the front can tap the backs of the men in the ring to hand them the buy and sell orders. I hunch near a desk on the second riser.

Ed, an analyst with a shaved head and a huge mint green headset, wanders over to see what I'm doing. I ask why the men are wearing safety glasses. "Some guy down there probably lost an eye." Then he's back on the phone.

An older guy in a loud coat with a lot of flags attached to it looks up from where he's huddled over his phone and offers this: "These guys are all serious when they have to be. But look out! When they're off a minute they'll put shaving cream in your shoe. Know what I mean?"

In the NYMEX I know nothing, except that the roar is giving me a pounding headache. I stare at the crowd of jumping, screaming men, the phone cords, the news bulletins from around the world scrolling on the walls, and I have no idea where to begin. It is if I have finally entered the central control room for the world economy and found it full of a nonsense of jumping rodents and random headlines generated by a squad of monkeys banging on typewriters. This is nothing like the careful world of spreadsheets, rigorously measured underground pressure, rigid schedules, and long-laid plans I've visited so far. The headache has reduced my thoughts to short thumps. What if. Oil prices are determined. By muscle twitches. Not logic?

"In the beginning they try to puke them out," a trader says. "The big players fake the mood for the day and try to get the small players to cave." The big players are refiners and producers, who often break up their buy and sell orders among different traders so no one can discern their strategy. It's now 10:07—just seven minutes after the starting bell. Puking and caving has brought the price for September crude down 8 cents, a fact I know only because it's on a digital readout on the wall.

The next 19 minutes and 31 seconds pass very slowly. The prices for September continue to fall, a penny or two at a time, and I try to calm my head by writing down each tick of the price: $44.32, $44.30, $44.26, then it falls another 5 cents, and then 2 cents and then 8 more.

At 10:25 the secretary of Homeland Security appears on the TV. The slide continues: $44.10, $44.05. And at 10:26 and 31 seconds it hits $44.00.

Seconds later a note appears on the TV screen above the pit: REPAIRS TO PIPELINES IN IRAQ! "Sell. Sell. Sell," screams the crowd. The volume of noise rises, the arms pump the air more vigorously. The men in the pit look like they're boiling in a cauldron, their bodies rising and falling by thermoconvection.

September crude continues falling to $43.97, when suddenly it turns around and heads madly upward fifty seconds later: $44.30. After another twenty seconds it's up another 10 cents. The price is now exactly where it started, but in less than an hour it has traveled 86 cents down and back up.

There is a lot of money to be made in these changes, even though ultimately they may not affect the actual oil price much. One criticism of the oil market is that it encourages speculation, which possibly drives up the price of oil. "People in the oil industry say that speculation is bad," says Professor Robert Weiner of George Washington University and Resources for the Future, "but the market needs speculators. They're like grease. They lubricate the system by taking sides of the deals. Without speculators, there would be no NYMEX because buyers and sellers wouldn't have anyone to sell to and buy from."

The important question, says Weiner, is whether speculators are smart money or dumb money. In other words, do they make informed decisions that make prices more accurate, or do they buy in a sort of mindless pack? He did a study of more than a million records of daily activities by hedge funds and other market speculators, and concluded that they were largely "smart money" and there was little sign that they were herding each other into frenzies of senseless buying and selling. Speculators may actually soften price shocks. Weiner studied trading during the first Gulf War, when prices shot up and fell back quickly. His research suggests that futures trading played a mitigating role at a time when the market was in crisis.

I ask Ed if he wants to be a trader. "It's risky to move into trading," he says. "You could lose your customers. All these guys are competitors." He gestures to clusters of analysts on the floor, men in khakis across the pit hunched over their phones. "They figure out who you're talking with and they call all your customers. They could be instant messaging my guys right now, and that's not good."

But even the analysts make a lot of money. "I work 9:45 to 2:30 and I make triple what my friends make," says Ed. In his off hours, he's a competitive bike racer.

"The value of money here? It's nothing," he continues. "At Christmas the brokers give $500 like it's nothing: It's the equivalent of $50 on the outside. Nothing." Nobody here looks rich. In fact, they look like they all go to a Catholic school where the uniform of chinos and sneakers covers up subtle differences in class.

Ed leans forward conspiratorially. "That guy in the blue? He's a multitrillionaire. I've seen him spend a quarter million at a charity auction for fun." He points at a guy in ill-fitting trousers, with a wrinkled shirt. "He's a multibillionaire." Ed leans in closer. "Nobody dresses flashy here but the watches are how you tell. They have a good day and they go downstairs to the jeweler and buy a really nice watch."

At 10:30 the U.S. government's Energy Information Agency releases its weekly report on the crude oil inventories held by refiners. Stocks are down, they say, and that sends prices up.

Inventories are one of the ways the United States influences the oil market, says Michelle Billig, an analyst at the consulting firm PIRA Energy Group. When U.S. refineries have full tanks, they can afford to wait for oil prices to fall before they have to refill them, so the inventories act like cushions against the shocks of the market. But over the last twenty years, as companies have merged and cut costs, they've reduced their stocks from four months' worth of crude to less than two. Futures prices tend to rise when that cushion is thin, as it is today. Unfortunately, the two other cushions—the SPR and the Saudis' ability to pump extra oil—have become problematic. Without the cushions, Billig says, "the market becomes very aware of risk."

Back on the floor, prices have climbed to $44.70 when Shell announces that it's shutting down 90,000 barrels of oil per day in Nigeria because of violence in the oil-producing areas. Prices rise to $44.95, as the market worries and starts buying. There's an expression here that "the most expensive barrel of oil is the one you don't have."

A red announcement arrives from Platts, an energy news service: "Saudis have 1.3 million barrels in spare capacity," meaning that the Saudis can and will produce more oil. In the next four minutes, prices fall a dollar, back to $43.93. One of the ironies of the free market in its current form is that it is dependent on a cartel with the ability to increase its production by 5 percent literally overnight. "Spare capacity only exists in countries that don't operate on market fundamentals," says Billig, "like OPEC."

There's another ironic twist: The Saudis themselves have started keeping a watch on inventories, the other cushion. "You have to watch that like a hawk," Saudi oil minister Ali Naimi said in 2004, explaining that when inventories fell, OPEC put more oil on the market, and that when they rose, OPEC pulled back, keeping prices higher. As a result, the United States has less influence over prices than in the past.

Let me recap: It is 11:20 in the morning. The price of crude for the month of September has traveled a total of 243 cents in one hour and twenty minutes, for a net loss of 47 cents a barrel. Not only that, the prices for crude in October, November, and December have also been traveling up and down on their own curves, all being broadcast through these fluttering hand gestures and croaking voices across the pit. It's not even noon and one trader has his hair, shirt, and tie askew, like Robert Mitchum after a bender.

"For us, short term is the last three minutes," explains one of the analysts standing by the side of the floor. She is slight, her blond hair is in an efficient clip, and she keeps a distant eye on the action in the pit, as if it's both fascinating and ridiculous, like Jell-O wrestling. Someone told me her water broke while she was on the trading floor and she stayed at work. I ask her about that. Yes, she says, she stayed until the close of trading and then took a taxi to the hospital. I expect her to be a stoic, but she turns out to have a deep, almost spiritual faith in the moment-to-momentness of the market. She speaks stream-of-consciousness as the trading heats up. "None of us know. This whole market, nobody knows where it's going. We might have opinions or experience, but all you can really expect is a mystery of faith." She says she's been in the market for twenty years and this five-year climb in prices is a "straight shot up. Unbelievable. Higher highs and higher lows. Mass psychology."

**The crude pit** is an intoxicating mix of operatic drama and Super Bowl spectacle, but I need more perspective than the play-by-play narrative I'm getting here. So I call Charley Maxwell. In the world of energy predictions, Maxwell is a lion. He started work in the oil business in the late 1950s and became an analyst in the 1960s, before oil was much on anyone's mind in the United States. In early 1973, before the Arab oil embargo, he is credited with issuing "Maxwell's Warning," when he told a group of Detroit executives that the oil world was about to tilt dramatically from cheap to expensive oil. He called it an "energy crisis," and no one listened. A year later he was famous.

Charley Maxwell is still analyzing the industry. He believes prices now tell a new story. "Up to now, the rises and falls in price have been approximately equal," he says, "In the '80s and '90s the price would go up to $40 and come down to the low teens, then go up again. So there was a sense that the energy crisis was rectified and you don't have to really change. There was a continuity to the ups and downs which made people think that up is followed by down. But this one isn't going down."

Maxwell says he began to watch the market closely in 1998, when it appeared that production from major non-OPEC oil exporters was starting to fall off. He ticks off the years when countries recorded their highest production: the United States in 1970, Egypt in 1996, Argentina in 1998, Colombia in 1999, the U.K.'s North Sea in 1999, Australia in 2000, Norway's North Sea in 2001, Oman in 2001, Yemen in 2002, and Mexico and China are near their peaks now. These countries are still producing, but, like the drillers in East Texas, they are going for oil that's no longer easy, or cheap to extract.

All of this adds up to a picture of the non-OPEC oil producers in decline, tilting the whole oil supply table back toward the OPEC countries. "You could argue that because OPEC countries no longer have to compete for markets against non-OPEC producers or even the Saudis, they suddenly don't feel the need to invest money to increase production," says Maxwell. "The less they invest, the faster the price will rise. A perverse application of economics." The economically irrational gasoline consumers of the United States are on a collision course with their opposite number—economically irrational oil producers.

Now Maxwell has issued another warning. It is dense, full of analyst-speak, but it is definite: The days of the Black Giant, the drama of oversupply, are over. "The new basis for doing business—that is, restrained supplies at higher prices versus virtually unlimited supplies at the lower prices of the past—will entirely change the structure of opportunities, supplier-customer relationships, the establishment and direction of energy-producing organizations, and the valuations we have historically put on them. After 145 years of [the] petroleum industry, a wholly new one is emerging. None of us has ever seen the like of it before."

**Ed comes back.** He asks if I think this place is a target for Al Qaeda. Aren't you supposed to be telling me that? I ask. No, he says, what do you think? I think that everyone who trades crude walks past "the Hole" on their way to work every morning. Once they're here they spend all day reacting to news about terrorist attacks around the world while wearing buttons and flags to remember people who died in the WTC. Perhaps the Hole acts the way black holes do, warping the shape of the space around it, folding the NYMEX into that warp. The rest of the country may have distracted itself from thinking about terrorist attacks, but they are a visceral muscle memory at the NYMEX and a virtual certainty in the market. Now that I'm here, the concept of the "terror premium"—a rumored extra charge of $5 or $10 per barrel in the price of oil—makes perfect sense.

One of the curious things about the exchange is that it efficiently aggregates information, but it doesn't make predictions, much less ruminate. On my second day on the floor, I notice a trader wearing a jacket covered with small photos of Che Guevara, à la Andy Warhol. I start laughing and whack Ed on the shoulder—look, Che Guevara in the heart of global capitalism! "You know who he is!" says Ed, calling the trader over excitedly. Apparently, no one on the floor today knew the famous image of Che, even though they have spent the last two years trying to second-guess Venezuelan president Hugo Chávez, a known admirer of Che.

Hugo Chávez controls the largest reserves of oil in the Western Hemisphere, and he's a big player in the future of American gasoline prices. He's been furiously reinventing the Venezuelan state oil company, which owns refineries in the United States and even supplies fourteen thousand CITGO stations. Recently, he's been using petro-diplomacy to create a Latin American trade block with a distinctly anti-Yankee bent. To understand Chávez's behavior and ambitions, you need a little sense of the history of leftist populists in Latin America, including Che. But at the moment Che remains an unknown on the trading floor.

Until recently, the NYMEX, like the United States, has been insulated from the outside world by cushions both real and psychological. For decades, special relationships with oil producers made the supply seem assured, and problems like terrorism seemed far away. Oil fields and their political, environmental, and economic troubles were remote. We've had the luxury of perceiving our relationships with oil suppliers as mainly economic, looking at their output more than their motivations. But all of this is changing, as the world gets smaller and what were once private political matters in oil-producing countries have started to affect the oil price. Since the United States has chosen—for the time being, at least—not to control oil demand, the history and psychologies of oil producers are becoming a personal concern for Americans every time we visit a gas station. Now it's our business to know what's on Chávez's mind. And if Maxwell's Warning is correct, we have left our old insulated world behind.

**Twenty-three minutes** before the market closes, September crude has fallen from $45.75 to $45.57. The crowd is screaming. In the pit, the traders are wiping their sweating faces on their sleeves, as if they've been working out. The price is bopping around $45.50. Vice President Cheney is on TV saying "A sensitive war will not destroy the evil men who killed three thousand Americans." Near my ear a woman shouts, "Are you coming over here to tell me it's going to close in my fucking face?" The traders are moving up and down as if the heat's been turned up higher. A small trader flies backward off the risers, squeezed out by the larger shoulders pumping the air near him. The cards are streaming overhead like seed pods in a windstorm. At two-thirty the horn blows and everything stops. The cards float toward the floor. The traders turn around.

The analysts pass around some alcohol and a tissue to clean their phones, which they say stink by the end of the day. On the floor there are time-stamped cards, paper clips, pink papers, gum wrappers, rubber bands, and a paper bag. The traders reach under the stands for their backpacks. I stop one, who talks to me while holding his maroon backpack as if I'm a teacher scolding him in detention. "Yeah. I traded, what . . . four thousand contracts today." I ask him how many barrels that is. "Ummmm. Four million." His voice is a squawk, but not as bad as the man with the Popeye voice. "That's one-fifth of the U.S.'s daily needs," I say. He looks embarrassed, tearing up some cardboard and throwing it on the floor. "It doesn't mean anything," says another trader. "Yeah. It's nothing but paper," says someone else.

**"Crude Oil Light Sweet"** is written across the top of Tom Bentz's graph of oil prices from 1983 to 2004. From the left side of the paper to the right, the line starts high and falls quickly until it rises on the other side, looking like a jump rope held between two schoolkids, except for a sharp price spike in the middle, for the first Gulf War. If you take Charley Maxwell's warning seriously, the last twenty years of low prices look like an anomaly rather than normal prices.

Reading the chart with Tom is a way of understanding why oil used to be cheap. Tom points to when OPEC lost control of the oil market. OPEC tries to keep prices high by allowing each producing country a quota of oil and forbidding them to export more than that. This works fine when demand for oil is strong, but between 1979 and 1983 consumers had cut worldwide oil demand from 52 to 46 million barrels a day, non-OPEC suppliers were going great guns, and OPEC's members were cheating madly on their quotas. Saudi Arabia watched its earnings fall from $119 billion in 1981 to $26 billion in 1985, and then decided to open its taps. Oil was at $31.75 when the Saudis started to pump, and then it began to fall.

Four months later, prices hit bottom. "It's kinda funny," Tom remembers. "For some reason that night [in March 1986] I'd had a dream that the prices would bottom out at $9.75. The next morning I was literally on the phone telling my friend on the floor if you see this price—$9.75— buy because that's the lowest it'll ever go." I ask him if he doesn't think that's a bit weird, having a dream that predicts the oil market. He shrugs. "Your body and mind get into the flow of prices and you can feel where the prices are going." I have no empirical evidence suggesting that Tom Bentz can feel the oil market in his bones, but no one could have predicted that increasing the world oil supply by merely 3 percent would cause a 70 percent drop in the price of a barrel.

For a cartel, OPEC has a hard time keeping its act together. One study of OPEC producers' actions (how much oil they actually produced, not what they said they produced) found that it was hard to distinguish when OPEC members were cooperating with each other from when they were competing. "OPEC is much more than a non-cooperative oligopoly, but much less than a frictionless cartel," concluded economist James Smith. And for most of the last twenty-five years, OPEC has struggled to keep any hold at all over prices.

Back to the chart: A spurt from $15 a barrel to $41 in 1990, when Saddam Hussein invaded Kuwait in August. "It was the most insane," remembers Tom. "The surprise was Iraq invades Kuwait and immediately 2.5 million barrels are off the market. That was viewed as a potential catastrophe. What's next? No one knew. Immediately the market exploded." In three months the price of a barrel nearly tripled. "Normal volatility was in the range of 30 cents," he says with amazement. "The mother of all bombs, what was that?"

Five months later the market was poised to go nuts again if the United States was going to go to war in Kuwait. "Everybody was glued to CNN," says Tom. "I had two phones on one ear and another phone to the trading floor on the other, and was literally watching [Secretary of State] James Baker walk into the room. It was kind of like that movie *Trading Places,* where the whole world was sitting and waiting for the news. When he said 'regret' I yelled 'buy 100,' and the market exploded. It went up $7 in a minute and people still wanted to buy. No one knew what the high of the war could be. People were talking about $100 a barrel. If bombs were flying in the Middle East, it could cut 10 million barrels a day. People had the idea that this could be the end of the supply as we know it." That night Tom kissed his children good night, wondering what their future would be.

The war was anticlimactic, and oil prices fell immediately. "By the morning, it seemed clear that we had gone in there and where was the big fight? The following morning, crude opened 9 dollars lower and *boom!* A lot of companies lost money. The psychology was that prices were going higher and then all of a sudden the psychology was: Wow, it's not as bad as we thought. Plus there was oil coming from the SPR and on top of that Saudi Arabia says they can put out 10 million barrels a day overnight to cushion prices." Crude drifted back to around $20 a barrel. The United States' willingness to go to war to protect the oil status quo made people feel that political problems in the Middle East were far away, perhaps even irrelevant.

Throughout the 1990s, prices sloshed gently up and down. When Asia's economies were doing well, they used more oil and prices went up. When they crashed, they brought on the next big low: 1998, $10.35. For consumers, complacency set in. OPEC seemed to have lost its power, and new oil was entering the market quickly from Russia and the former republics. Some analysts gleefully danced on OPEC's grave, forecasting that oil would soon be $7 a barrel and mock-fretting about a "crisis of cheap oil." Congress even let the authorization for the SPR expire for months in 2000.

For oil producers, this period was a disaster, and a lesson. "In 1998 alone OPEC members lost something in the region of $60 billion," OPEC secretary general Ali Rodríguez Araque explained later. "That came as a cold, hard shock to our governments, who were forced to cut budgets and slash development spending. But it also acted as a sharp reminder as to how complacency can be one step away from a fall. That crisis pushed us to the very limits of the pain barrier."

OPEC countries weren't the only ones feeling the hurt. In 1999 two non-OPEC countries, Russia and Mexico, teamed up with Saudi Arabia and Venezuela, cutting production to push oil prices back into the $20 range.

Back to the graph. On September 11, 2001, Tom was doing a live phone interview as he watched the second plane hit the tower of the World Trade Center on CNN. "My voice cracked," he says. "At that moment we knew it wasn't an accident." Oil markets dove.

Soon Tom developed a bad cough. A doctor told him it could be a "WTC cough" from burning debris, but it persisted. In November he was diagnosed with a tumor the size of a grapefruit hanging off his aorta. "It was basically choking me," he says. "I had palpitations for years but I just thought it was stress. You know, you're dealing with the markets every day and the flow going up and down and there's a lot of stress and tension every day. I just thought, you know, I'm out of shape."

The tumor was inoperable, and his only recourse was a very severe course of chemotherapy. He continued to go to his office until the chemo got the better of him. As oil prices started an upward climb from a low around $18, Tom's weight fell from 210 pounds to 130. "I was practically dead." Amazingly, the tumor shrank to the size of a walnut, and his surgeon scraped it out.

When Tom came back to work in 2002, the U.S. economy was recovering. China's economy was continuing to grow at 9 percent per year, requiring huge amounts of energy. As their per capita income rose, China's wealthy started buying cars. Demand began to snowball. Oil imports went up 15 percent in a year. China's factories were going nuts! There was even a rumor that China now made 70 percent of the world's ceramic toilets, and kilns require lots of energy. Asia was burning an extra million barrels of oil a day just making products that got shipped to the United States.

In late 2002 the Venezuelan strike sent prices higher, and then the United States invaded Iraq, sending prices still higher. This time Saudi Arabia couldn't, or wouldn't, step in to lower the price. The Saudis had less oil in reserve, but their attitude had changed. "After 9/11, Saudi Arabia realized that trying to please the U.S. didn't have much impact. And then they were displeased by the U.S. decision to invade Iraq. So they could go ahead and allow for higher oil prices because there will be neither political nor economic repercussions," says energy economist Herman Franssen.

And without cushions, or brakes on consumer demand, the oil market is becoming the Wild West, a geopolitical theater where tensions and fears of all sorts are acted out through oil prices. A month after my visit to the trading floor, in September 2004, oil will crash through the next plywood barrier of $50 on the news that a freelance militia leader in Nigeria's oil region has threatened to take on the government and attack foreign oil companies. The globalization that oil has been fostering all these years is catching up with itself.

When prices hit $75 a barrel after Hurricane Katrina wiped out Gulf oil platforms and refineries in the fall of 2005, fingers pointed in all directions: OPEC blamed lack of refinery capacity in the United States. Congress blamed the oil companies for "obscene profits" and China for consuming more oil. The oil companies blamed speculators in the market and Detroit for making SUVs. And some people blamed OPEC, terrorists, and on and on. It was the classic blame game. But consumers had long ago lost control of the oil market, and demand had been rising for years.

By all rights, the world should now be in an "oil shock." One analysis said that every $10 rise in the price of a barrel of oil cost the world

$255 billion in GDP. Every major recession but one was preceded by a sharp rise in oil prices. But at the moment the world economy is galloping along and gobbling still more oil even though consumers are paying trillions of dollars more for oil. The secret ingredient is credit—just as consumers are putting gas purchases on credit cards, on a macrolevel the United States is financing oil purchases on credit, insulating the economy (for now) against the shock of high oil prices. But the future of oil is being made elsewhere, in Venezuela, West Africa, and the Middle East.

Before I leave, I ask Tom if the tumor had changed how he felt about the oil market. He says life and the market are one and the same. "You don't know what tomorrow will bring. Every day you do your research and strategize about your next move. Just when you think you have a handle on things, something turns your plans upside-down. When that happens you realize you know nothing."

# 7 VENEZUELA *NINETY YEARS OF "THE STRUGGLE"*

I arrive in Caracas, the capital of Venezuela, late one night in the spring of 2004. Riding into the city from the airport, the hills look like they've been drizzled with glowing lava, but the twinkling lights are really the disorganized barrios that cling to the hilltops. In Caracas, great numbers of poor teeter in the crumbling hills while the middle class and rich live in well-guarded high-rises in the valleys. But this system is in flux, and as traffic slows to a stop on the freeway, the taxi driver wonders if there's been another riot. People say that Venezuela is on the brink of civil war, a surprise in a country with the largest oil reserves outside of the Middle East and nearly fifty years of history as Latin America's oldest, most stable democracy.

Forewarned of armed robberies by Venezuelan expats living in the United States, I'm carrying all of my money in my sneakers, hundred-dollar bills stuffed under the insoles. The bills—enough to support approximately three hill dwellers for a year—give a floaty, not-quite-here spring to my steps. Initially I credited this unmoored feeling to the money, and then to the cushioning, but I think now that it was the result of a different gravity. Traveling from a giant oil consumer like the United States to an oil producer like Venezuela is like falling down a rabbit hole: Here the familiar features of oil in the United States—its economics and politics—have been turned upside down by the fundamental contradictions about sovereignty, power, and national mission that lie in the heart of the petrostate. Soon after my passport is stamped, I am hit by the implications of the new gravity: The budget of the Venezuelan state is $26 billion (in 2004) while the income of its national oil company PDVSA (pronounced pe-dah-vay-SAH) is $42 billion. Venezuela is, in effect, a subsidiary of its own oil company.

**The next morning** I buy a pastry at a café in the lovely upper-middle-class neighborhood of Altamira. Over my head a TV tuned to an opposition station blares ads of a young woman undulating in an orange dress, shouting anti-Chávez slogans. Populist Hugo Chávez was elected in 1998 but has battled to hold on to power since then. The struggle says a lot about the contradictions of life in an oil state. The park in front of the café comes into focus: In the center is a lavish impromptu shrine to the Virgin Mary, tended by matrons who oppose the president. Nearby storefronts are boarded up, either left over from the last riot or in anticipation of the next. On the edges of the park there is cryptic graffiti: RR O 350, which advocates war against Chávez by referring to the part of the Venezuelan constitution that permits armed rebellion, which happens to have been written by Chávez himself. It is only the first of many ironies. The next morning an effigy hangs from a tree in the park. A sign around its neck compares Chávez to Hitler.

The president's strongest supporters are the poor from the hills, who flooded the downtown streets in 1998 shouting "We are hungry." In 2002 they came out again to support Chávez after he was removed in a short-lived coup attempt. The president was out for less than two days, but that was time enough for the United States to publicly endorse his overthrow by an undemocratic government that quickly dissolved parliament and the judiciary. Chávez came back to power—thanks to his crowds—and now treats the U.S. government as an enemy. When I take a taxi out of the city center, toward the bleached-out hills where the ranchos are, the graffiti changes: YANKE GO HOME, GRINGOS FUERA, and AGAINST IMPERIALISM THE PEOPLE ARE ARMED.

For nearly ninety years Venezuela has been the United States' unquestioned sidekick in petroleum. Venezuela sells 70 percent of its oil to the United States. Oil money, in return, makes up a third of Venezuela's GDP, half the government's budget. In Venezuela, the relationship of mutual dependency with the United States is omnipresent and filled with tension. "There is this fixation on the U.S." says Maruja Tarre, a professor of oil politics at Simón Bolívar University. "Everyday life in Venezuela is America. Even without an invasion, the U.S. influence is stronger because of oil, baseball, McDonald's, Mickey Mouse. And the U.S. is paradise so there is hate too."

The United States has taken Venezuela's contributions for granted, when it remembers them at all. Venezuela, five days' tanker ride to the south, supplies 12 percent of the oil and gasoline the United States uses. Americans are more dependent on Venezuela than on Saudi Arabia, because Venezuela's big fields are so close and their integration with the economy so seamless. PDVSA has insinuated itself deep into the U.S. heartland, where it owns eight refineries and supplies a network of thousands of CITGO gas stations. The Venezuelan ambassador to the United States has even described his country as the United States' "domestic" oil supplier, fully aware of the contradiction of claiming sovereignty from U.S. intervention on one hand and claiming full integration with the U.S. economy on the other. His North American listeners took comfort in the statement, assuming it meant that more cheap oil was on the way, but the ambassador knew it was much more complicated than that.

In the winter of 2003 the United States was taken completely by surprise when Venezuela's oil production fell suddenly from 3 million barrels a day to almost nothing. Anti-Chávez oil workers had gone on strike in another attempt to oust the president, but the United States was oblivious because it was preparing to invade Iraq. No one in the United States could imagine that Chávez would survive the strike, or that Venezuela would fail to send its oil northward, as it had every day for decades. Unwatched, oil prices began their climb from $26 a barrel toward $40 and beyond as 200 million barrels of Venezuelan crude failed to come to market over the next three months. "Analysts . . . underestimated President Chávez's willingness to sacrifice oil to politics," writes oil security analyst Michelle Billig. It was a bit of American myopia, a tendency to see oil as a purely economic substance, an even exchange of a commodity for money.

**Around Caracas,** signs have appeared with the words PDVSA PARA TODOS—the oil company for everyone. For most of its existence, PDVSA has been the oil company for hardly anyone, because it employs fewer than fifty thousand people out of the country's 26 million citizens. Some of the new signs feature a picture of an oil derrick, others the face of Simón Bolívar, who liberated Latin America from Spain in 1811. Oil is to be Venezuela's new Bolívar, liberating the country from poverty and American influence. Chávez is only the latest in a long line of leaders who've made similar promises, and every one of his predecessors has failed, tripped up by the inherently tricky nature of oil itself.

One morning I go to Plaza Bolívar, in the very heart of Caracas, where a statue of the Liberator stands above the cobblestone square. My guidebook describes this place as the holy center of Venezuelan nationalism. "If you find yourself in Plaza Bolívar," says the book, "respect must be extended to Bolívar's statue. Don't cross in front carrying bulky packages, or loiter in beach clothes."

The guidebook must have been written in a different time, because the Plaza Bolívar, like Chávez's Bolivarian revolution, has an improvisational air. At the Liberator's feet stands an androgynous Chaplin mime. Nearby a toothless man grabs at passersby. "You little fascist angel," he shouts at an infant who makes eye contact. And then there is the drunken woman in a red beret and smeared matching lipstick shaking a large wrench. Suzana, who's translating, whacks me. "That one we call Commandante Pliers."

Whatever Plaza Bolívar was before, it has changed. Hulking military vehicles crouch among the baroque buildings. Peddlers squat nearby selling pictures of the saints: Bolívar, Christ, Che Guevara, and Chávez. Nearby is a monument to the martyrs of petroleum: An effigy of an oil derrick, painted in the colors of the Venezuelan flag, is placed not far from the plaza, in honor of the Chávez supporters who were killed during the unsuccessful coup attempt of April 2002.

A young woman, terribly thin and all in black, with shellacked dyed blond hair, approaches and presents herself to Suzana. "Oil is the independence, the sovereignty of our country," she says, her concave chest filling with air. Her name is Gabrielle, and she has dedicated herself to petroleum, perhaps as people once symbolically defended Bolívar's dignity in the square.

Gabrielle says she protected the national patrimony during the strike. "PDVSA was the property of the Venezuelan people until they [the strikers] ruined it. They tried to break the country and sell the oil to other nations, but we took the company and slept there every day. Now the people own PDVSA." During this speech Gabrielle has become so large I see nothing but her angry face and thin arms. She has grabbed on to the current of anger and revenge, the sense of oil stolen—ideas much older and more substantial than herself—and used them to inflate her small body. Upon filling herself with more breath, she says that she's willing to fight for the oil again.

A friend wearing a fishnet top with push-up bra joins her, declares that she's also a member of an armed revolutionary group, and heads off to send an express package. In the new Plaza Bolívar, where the talk is of oil and war, swimwear and packages are unobjectionable.

"Right now everybody's a guerrilla," says Suzana sarcastically as we leave the plaza.

Suzana takes me over to the former PDVSA headquarters, an aloof ice cube of a high-rise in steamy Caracas. After the strike, Chávez set out to both dominate the oil company and demonstrate that he was ripping apart its rigid old hierarchy for the good of everyone, particularly the poor. The first symbolic casualty of the transformation was PDVSA's headquarters, which became a "Bolivarian University" for poor students. Other transformations followed. In 2004 PDVSA spent $1.7 billion on a literacy program, agricultural development, and low-income housing. Chávez also began use the country's oil to redefine Venezuela's role in the world, starting by trading it to Fidel Castro in exchange for Cuban doctors, who set up clinics in the poor ranchos in the hills.

Today there is a lot of activity at the headquarters-cum-university, but not much going on. Students are milling about, lounging in the cafeteria, and talking in gaggles on the landscaped lawn, but there are no classes and few classrooms. An area they are calling a clinic has no equipment and no files, just blue drapes over a few chairs. Construction workers are running around, as are student guards carrying walkie-talkies. An earnest and excited student says he and his classmates are designing their own courses. He's planning to study environmental management. "Right now we're totally dependent on oil. Chávez says we should sow the oil."

This is an idea that dates back to 1936, when novelist Arturo Uslar Pietri wrote an influential editorial saying that Venezuela needed to "sow the oil" to develop the country. Buying development with petrodollars has been a recurring rallying cry ever since, although the oil money has chronically failed to deliver. What was overlooked, perhaps, is another part of Pietri's editorial, which says "the destructive [oil] economy is one which sacrifices the future in favor of the present. Taking things into the realm of fable writers—it is more like the cicada than like the ants."

In Venezuela the cicada—with its cycles of boom and bust—has prevailed over the industrious ant. Between 1980 and 1999, Venezuela's income fell by a quarter, and nearly a fifth of the country slipped below the poverty line. More than half the population now lives on less than $2 a day. Decades of talk of using oil as a seed to be sown without results has led to a suspicion that oil makes the country into nothing more than an "oil factory" for the United States. So when oil is touted as the substance that makes Venezuela the Bolívar of the world, the liberator of the masses, there is always someone in the background muttering (equally convincingly) that oil is the "devil's excrement," an inescapable curse.

**Venezuelans became** the people of petroleum by a coincidence of history, geology, and corporate interests. The country's oil arrived when a gusher of 100,000 barrels a day shot into the skies near Lake Maracaibo in 1921. It showed up just when U.S. and European oil companies were pulling out of Mexico and Russia to punish their revolutionary governments. Although local farmers viewed the gusher as a curse on their fields, multinational oil companies saw the oil as a replacement for what they'd lost to socialism.

According to Spanish law, the oil in Maracaibo belonged to the farmers who owned the land above. U.S. oil companies didn't want to negotiate with the farmers, perhaps because the issue of land claims was a mess in Texas and Pennsylvania, and perhaps because they didn't want the farmers to become an alternative center of power, so they proposed that all oil rights lay in the hands of the state.

The Venezuelan state then consisted of the dictator Juan Vicente Gómez, a nearly illiterate farmer whom the U.S. government had helped gain power. He didn't think of the state as a nation with citizens, but more as one of his farms, with resources and inhabitants to dispose of at his will, and he didn't object to taking possession of the oil underneath them either. And so modern Venezuela was born as a triangle of state and foreign oil company, with citizens as an anxious third. In 1922 the three U.S. oil companies themselves wrote Venezuela's hydrocarbon law.

"It is one of the great ironies of history that foreign oil companies, the epitome of private enterprise, are largely responsible for the etatism characterizing Venezuelan development," writes Terry Lynn Karl in her classic study of why oil producers are the way they are, called *The Para-*

*dox of Plenty: Oil Booms and Petro-States*. In trying to punish the socialist states of Mexico and Russia, the companies invented a parallel world, a kind of one-sided socialism, where the state's main role was interfacing with the foreign oil companies and distributing oil money to the population. In very general terms, the state of Venezuela was designed as a sort of electrical wall outlet to supply energy to an American cord, which carried it northward to the United States.

The oil state gave the United States a way to resolve its need for international business relationships with its deep isolationist beliefs. In Venezuela, the United States was not interested in being a colonial power, but it wanted access to oil and rights and security for U.S. companies. Central to this arrangement was a faith in self-interested benevolence rooted in the idea that "development" would occur through sensible central government spending, thus making oil exports a net benefit to the population. The United States saw Venezuela as partly its own creation. "If there had been no perfect illustration of what U.S. technical and capital resources could do for the world's underdeveloped areas, it would have been necessary to invent the Republic of Venezuela," wrote *Fortune* magazine in the 1940s. By the early 1950s the National Security Council went further, saying that "oil operations are, for all practical purposes, instruments of our foreign policy."

Once conceived as a convenience to consumers, and an arrangement that distinguished the United States from the colonial powers in Europe, the petrostate has matured into a hazard itself. Lurking within it were instability, poverty, nationalism, and deep anti-American feelings. The 2001 National Energy Policy, written after secret consultations between Vice President Dick Cheney and oil executives, concluded as much. "America 20 years from now will import two out of every three barrels of oil—a condition of increased dependency on foreign powers that do not always have America's best interests at heart." This statement didn't get much attention in the United States, perhaps because it was bland and obvious, but in oil-producing countries it caused a lot of concern, particularly after the U.S. invasion of Iraq. In Venezuela, many people interpret it as a virtual declaration of war.

After we leave the Plaza Bolívar, Suzana sits on the steps of a barber shop to change her infant son's diaper. I ask her what she thinks about the talk of violence. "War is better than being taken over by the Americans who want our oil," she says brightly.

Its destiny determined by oil and the United States, Venezuela is a philosophical leader among world petrostates. Since the 1940s, Venezuela's intellectuals have applied themselves to perfecting the relationship with their biggest customer, and they have done it superbly, striking a savvy middle ground while increasing Venezuela's share of the profits. It was Venezuelans who led the movement to start OPEC in the late 1950s. After the oil company nationalized in 1976, PDVSA considered itself the equal of the major multinationals—a cut above other national oil companies. In 1999 Venezuela was instrumental in reinvigorating OPEC by getting two non-OPEC producers, Mexico and Russia, to cooperate in reducing production to raise prices. Under Chávez, Venezuela's ideas about national ownership of resources have been influencing oil-producing countries from Bolivia to Iran.

**Venezuela developed** by shopping spree. One of the country's most notable shoppers was military dictator Marcos Pérez Jiménez, who came to power in a *golpecito,* or little coup, in 1952, when two generals and an advisor met in a bathroom and decided to overturn the results of the recent election, with the private support of the U.S. ambassador. Between 1948 and 1957, oil put $7 billion in the hands of a government with a population of only 5 million, allowing Pérez Jiménez to spend an amount of money greater than the entire cumulative income of the country since it was colonized by Spain. He spent madly: freeways, boulevards, high-rises, luxurious country clubs, a modernist university, an aerial tramway, a mountaintop ice rink in the tropics.

Money was just waiting around to be taken. People compared the government to an enormous cow, giving milk to anyone who was in a position to catch it. The country began to import more and more and produce less, a typical symptom of Dutch disease, where resource-rich countries see other parts of their economies wither. (Venezuela actually had Dutch disease before the Dutch, but that term wouldn't be invented until the natural gas boom in the Netherlands in the 1960s torpedoed the country's economy. The condition should be called the Caracas cramp.) Imports rose while domestic production fell; and in 1952 Venezuelans spent $5.7 million importing eggs from the United States, though most of the population lived in the countryside, presumably perfect for raising chickens and eggs. Without a reason to invest in manufacturing to create jobs, money was spent on transient things, with the expectation that more would come from the oil. Beer made up half the increase in industrial production during the late 1940s.

In the countryside, a great migration started as people heard that Caracas was "as magnificent as the outlet of heaven." They headed into the city, hoping to catch some of the milk there, but most found a precarious existence on the shifting cliffs, where they built their homes in ranchos.

American officials were enthusiastic about their man in Caracas. (And the CIA World Factbook's profile of Venezuela still refers to "benevolent military strongmen," which seems anachronistic.) When he visited Caracas in 1954, U.S. treasury secretary George M. Humphrey enthused, "Here two and two make twenty-two instead of only four." But in 1958, while Pérez Jímenez sat playing dominoes, a riot in one of the new ranchos put an end to his rule.

After 1958, Venezuela became a democracy, but it continued Pérez Jiménez's precedent: government as enormous cow, distributing milk across the land. Democracy in Venezuela was based on a pact between two centrist parties to alternate power, paying for party allegiance with oil money. They gave jobs and basketball courts, houses, factories, and roads. Dozens of men were rumored to be on the payroll just to run the elevator at the central hospital. The vast amounts of money the parties distributed made it worthwhile for 96 percent of the country to join a party. One in seven people was clever enough to belong to both.

And nobody paid taxes. If you're an oil state, it's far more efficient to ask oil buyers for more money than to collect taxes from your population, which requires a vast network of tax collectors, a bureaucracy, laws that are fair, and a justice system to administer them. Collecting oil money, by contrast, requires a small cadre of intellectuals to set policy and diplomats to make it happen. So over the years, the oil state has evolved into something like a one-armed bodybuilder: The arm that faces the United States is rippling with muscles, the perfectly articulated fingers a work of art, but there is no corresponding relationship (or arm) between the state and the people of Venezuela. The political, economic, and psychological ramifications of this shape are profound.

"Systematically the government went after oil money rather than raising taxes," says economist Francisco Monaldi. "There is no taxation and therefore no representation here. The state here is extremely autonomous." Whether it's a dictatorship, a democracy, or something in between, the state's only patron is the oil industry, and all of its attention is focused outward. What's more, the state owes nothing more than promises to the people of Venezuela, because they have so little leverage on the state's income.

When a state develops the ability to collect taxes, the bureaucracy and mechanisms it creates are expensive. They perpetuate their existence by diligently collecting as much money as possible and encouraging the growth of a private economy to collect taxes from. A strong private economy, so the thinking goes, creates a strong civil society, fostering other centers of power that keep the state in check. Like other intellectuals I talk with in other oil states, Monaldi finds taxes more interesting and more useful than abstract ideas about democracy and ballot boxes. Taxes aren't democracy, but they seem to connect taxpayers and government in a way that has democratizing effects. Studies by Michael L. Ross at UCLA found that taxes alone don't foster accountability, but the relationship of taxes to government services creates a struggle for value between the state and citizens, which is some kind of accountability.

Venezuela might have discussed starting a tax system in the 1960s or early 1970s, when the country's oil output began to fall. But just when new sources of revenue seemed necessary, oil prices shot up during 1973's Arab oil embargo. Money started pouring in. For the rich, there were galas with cheeses the size of Volkswagens. For the middle class, there were trips to Miami, where they became known as Gimmetwos because they bought two of everything.

The state was also buying two of everything, on the way to building a Grand Venezuela. It began borrowing, assuming that money would continue to pour in from oil. When the price of oil fell in the 1980s, the state didn't reduce spending. With the downturn, some government departments even increased their spending. Venezuela borrowed money against its oil reserves, and its debts grew. In 1975 the national debt was $1.5 billion; by 1985 it was $26 billion, and by 1986 it was $32.7 billion, when debt payments were eating 40 percent of oil earnings.

In 1988 banks threatened to cut off loans if Venezuela didn't adopt the austerity plan of the International Monetary Fund (IMF). On the streets, the plan was called "being hit by a package" (*el paquetazo*) because it was a shocking break from the generous government of the past. The package restructured the country's economic relationships with foreign countries, destroyed the government's ability to provide jobs and favors, and (most important to the poor, whose numbers were growing) also increased prices on food, transportation, and utilities and doubled the price of gasoline.

The whole system of petropeace officially broke when an increase in bus fares set off huge riots in Caracas in early 1989. Government troops responded with arrests and killings. (The exact number of dead is not known, but it may be as many as a thousand.) The horrors of "the package" were associated with the United States and the IMF—just another coup. Later, when Hugo Chávez condemned "savage neoliberalism," Venezuelans knew exactly what he meant.

**Living in** the shadow of oil wealth has profound psychological and political effects. One day I traveled to one of the highest, most crumbly settlements in the ranchos to meet Yvonne, a thin, worn woman in a pale peach top, who has lived in a temporary shelter for twenty-seven years. Yvonne's glasses have been taped and one of her few remaining teeth has migrated to the center of her upper gums, which gives her otherwise desolate presentation a zany edge. In 1974, Yvonne was a kindergarten teacher with a two-story house when the government demolished it and promised her and her family a new one. She was relocated to a cluster of cardboard homes on this ledge and told she'd have a new house elsewhere in four months. "I expected to build a larger house and raise my children in a better place," she says, "but when I arrived here it seemed like everything was gone." Yvonne's future narrowed to this small dark shed where "even staying clean is a struggle."

Residents of the shelters have no toilets, so every morning they squat over newspapers, wrap their feces into a package, and then walk to the cliff, where they drop the package over the edge. Charlie Hardy, a former U.S. priest who lived on the ledge for eight years and brought me to meet Yvonne, calls the morning walk "a moment of ignominy."

For Yvonne, that moment has never ended. She had more kids, she lost her job, she raised children abandoned by other mothers. As time went on, she replaced the shed's cardboard walls with scraps of tin and wood and settled into a shell of disappointed expectations.

As I stand with Yvonne, I worry about her thin arms and her helplessness. A reporter who zips from one place to the next (in money-cushioned shoes) has to make a very big leap to understand the passivity that leaves a person waiting on a ledge for decades. Yvonne is obviously educated, informed, and hardworking. She is not waiting for a handout so much as she's permanently expectant and perhaps afraid that if she found her family a new spot, they'd miss what was coming to them, sooner or later. Her faith in the system is so great, and so tragically unfulfilled, that it is almost a religion.

And as I stand on the ledge I get a feeling of déjà-vu, back to a conversation I had with a former Miss Venezuela who now lives in Houston. She had an elegant voice, and even over the phone I felt I could smell perfume and thick carpet, but she began to talk about growing up rich in Venezuela. "People lived well in the 1970s, but there wasn't a sense of living in society. I took it for granted. You took as much as you could and gave back as little as possible." Her voice had a tinge of self-pity that is too easy to mock, but in retrospect I feel for her. "As individuals we were part of a culture based on either having wealth or no wealth. It wasn't based on being a contributor but on being a taker." Chávez's election, she says, was "an awakening" for her class that if they didn't take part in politics, "someone else would make decisions for me."

How did this trap of helplessness come to affect both the rich and the poor? Alfredo Keller has been polling Venezuelans for decades. He invites me to his office, which is in a mall in a nice neighborhood, to show me some of his statistics. Keller is in his fifties, with a natty European style and an office decorated with antique fire hats. He likes to pose questions with a bark and answer them with a crow. "Ninety percent of Venezuelans believe Venezuela is one of the richest countries on earth. Seventy-five percent say they are poor. Half see no way to leave poverty. What is that?" He pauses for a moment. "A social psychologist calls that an external locus of control. The world is so strong and I am so weak. This is all related to the cult of oil."

To Yvonne, the control lies with the people who got the money. "Practically, that oil money never came to the poor people. It went to those in power and stayed for forty years," she says. A neighbor of hers is more direct. "I came to the conclusion that someone had stolen the oil," she says. Who? "People from within and without. The oil has left the country unaccounted for. How is it possible we should have a rich country, selling oil to the U.S., and yet the condition of our schools, the bathrooms are subhuman?" An English language rap song outside her window plays "give 'em a kick in the ass."

Keller finds corruption a bit amusing—if you leave out the moral judgment, he says, it's just a distribution system that allows the state to buy loyalty. "We discuss it morally, but it's really an opportunity. It's a democracy of distribution. And if a country is not a matter of law, then you become rich through connections." In a sense, most Venezuelans agree with him, because twice as many believe that they will get wealthy through connections and loyalty rather than through productivity or institutions.

Fatally, nearly 80 percent of the country believes they will be rich if only they eliminate corruption—meaning that it's not corruption they oppose, but who's getting the money. "Every politician who has won has won on a speech to get rid of corruption," says Keller. "People buy these messianic promises that are impossible to deliver. And the political consequences are cycles of expectation and frustration." By the time the system broke down in the 1990s, three-quarters of the country believed that all previous governments were a disaster. They wanted change. They also wanted revenge against the people who took their money. "Now do you understand this revolution?" Keller roars. "People feel Chávez is right when he says they stole the wealth of the people, and we must punish them."

Yvonne voted for Chávez, but she remains ambivalent about him. "I thought he'd move the country forward and get rid of corruption," she says, "but they've kept his hands tied." She means the opposition, and the coup, and the recall vote. When Chávez comes on her TV in the afternoon to make one of his stream-of-consciousness addresses, she leaves the room.

Chávez is wearing a quasi-military outfit and he is bouncing boisterously about the podium. "Mass media has become a problem of public health," he says, "not only by consumerism but also traffic accidents that come from capitalism." He acts out a car commercial he once saw that showed a beautiful woman's legs. "You could see her legs. She opened the door. And you could see more leg! They said the car was for 'the leaders of the future,' and I criticized that." Leaders, he says, should not drive cars. That is American propaganda.

Watching Chávez is like watching a favorite drunken uncle. He is utterly charming and ingratiating, by turns garrulous and fanciful. He makes promises; he explains; he blames. He makes notes with a marker on pieces of paper, then folds the papers and puts them in his pocket. He shakes the constitution. "We are going against a thousand devils and coups." Keller describes the appeal of the godlike ruler in the oil state as a sort of existential crisis, "to be free in my dependency I need strong leaders." I feel dizzy and claustrophobic though I don't know whether it's Yvonne's dark shanty or Chávez's speech.

Chávez is an artist of the external locus of control, alternately debunking it by exhorting Venezuelans to do for themselves, to create co-ops and Bolivarian circles, and exploiting it by saying that the power lies outside, in those "thousand devils and coups." He is the ultimate strong leader, and his solution to the lack of connection between the state and the electorate in Venezuela is those markers he keeps in his pocket. Everywhere I go in Venezuela, I meet people who've sent him letters, because they see Chávez as the only man who can get things done.

I go outside to find Yvonne, who has changed into a blue dress so old it is becoming transparent. She is standing in front of a small collection of empty ketchup bottles and what appears to be some chocolate milk in a Fanta bottle. This is her kitchen—there is one broken chair and one whole chair.

She looks even more wounded, pulling her thin scabbed arms toward her chest like withered wings. "He's *bruto*," she says, "like a *burro*. He is too blunt. But he has some good ideas. He talks and talks without a point . . . and that makes me worry." Her nightmare, she says, is civil war.

But Chávez has taken some of the sharp edges off her life. A Cuban doctor has arrived at a clinic near the ledge and has set about treating diarrhea, asthma, and hypertension. Courtesy of PDVSA. A program called Mission Rivas, which is teaching anyone who wants to read, is going full force and the president is photographed hugging newly literate grandmas right and left. Also paid for by the oil company. The army is selling discounted frozen chickens, dehydrated milk, and rice, the packages decorated with cartoons illustrating the rights guaranteed in Chávez's constitution. And around the rancho, the people who are studying or forming Bolivarian circles say they finally feel like they have some control over their destinies. Listening to them, I feel wary of my own emotions because I so thoroughly want them to find comfort and self-determination. Is the revolution really just another scheme to distribute the milk, even if it's dehydrated this time, and going to the very poor?

People who oppose Chávez often dismiss his revolution as a flash in the pan, based on false hopes foisted on a gullible, poorly educated, long-ignored public. "These people were invisible before," says the former Miss Venezuela, "but this man discovered them. Now they feel they're part of an elite." She says that the poor think that the president is offering them a box of chocolates; they don't realize he's just offering them an empty box.

I get the sense that the poor know exactly what they are getting. "Venezuela is rich in oil, iron, and Miss Universe," a man named Javier says sarcastically as he leads me into his flimsy government-built tract home to show me the government-subsidized food in his cupboards. He spent his childhood in a shelter near Yvonne's, he says, and watched the army kill his friends during the riots of 1989. For Javier that was the end of any faith at all in the old government or the illusion of democracy. He sees government as a predator, and he's happy this one isn't eating him. He doesn't care that the economy has contracted by 20 percent under Chávez, because for the first time he feels like he's getting some of the benefits of being Venezuelan. "I've had to switch from whiskey to beer," he says dismissively, adding that he'd happily take up a gun to defend the president. He figures the nearby army barracks will give him one, should the need arise. (In 2005, as oil prices rose, the Venezuelan economy expanded dramatically. Chávez bought 100,000 AK-47s from Russia, reportedly for citizen militias to defend the nation.)

Javier's brother Tony adds: "Chávez is our son of a bitch. He's nationalistic, ignorant, but with big balls. Chávez doesn't have a plan, but he has balls, and we think he'll find a plan in the process."

**Oil has** always been Venezuela's main nationalist project. For generations of intellectuals, the immediate task has been figuring out how to get more money for the country's petroleum exports. They have coached that misshapen bodybuilder until his arm is an extraordinary revenue-extracting machine, a model for other oil states. The most influential of these nationalists was Juan Pablo Pérez Alfonzo, who was possibly the world's greatest philosopher of oil as well as the "father" of OPEC.

Pérez Alfonzo wrote the first radical restructuring of the contract between a petrostate and the oil companies in the 1940s, so that Venezuela got half of the profits from its oil. When the multinationals tried to punish Venezuela for the change, he convinced other oil states to insist upon the same fifty-fifty agreement.

During the 1950s Pérez Alfonzo lived in exile in the United States. He spent his time at the Library of Congress reading about the Black Giant of Texas, examining how the Railroad Commission's system of shutting in or limiting oil production worked. He felt that with the world's petrostates pumping full out, they were competing with each other and creating a glut of oil, rock-bottom prices, and a recipe for waste. Pérez Alfonzo didn't like waste. He returned to Caracas in 1958 carrying the germ of an idea about uniting oil producers. That germ became OPEC, the fractious cartel that controls the majority of the world's oil reserves.

"The first OPEC meeting was held in our house," says Professor Tarre. I went to visit her in one of the high-rises, threading through layers of security and elevators, reflecting that I was surely getting the bends traveling between the homes of the well-off and the poor. Tarre has photographs of the first meetings, and her face is enraptured as she talks about them. "Sheikh Tariki [of Saudi Arabia] was very good-looking." She shows a photo of the next meeting in Baghdad, where twelve men sit in a small room with a white tablecloth and a phone on a chair. The following meeting was more official, and next even more official as the group gained power. The early members were poets and aristocrats, but soon the countries developed intellectual bureaucracies, including oil ministries where lawyers and technocrats shared knowledge. As Tarre grew up and went to school abroad, she and her colleagues, as she puts it, "fantasized very much about OPEC" as a route to Venezuelan power and as a way of extending Venezuela's influence in the oil market long after the Middle East had surpassed its output.

As a way of maximizing oil income, OPEC was a brilliant move, but by the time of the 1973 Arab oil embargo, Pérez Alfonzo was disgusted with it. "He thought it was too much luxury when the Venezuelan minister of mines gave the [Saudi] sheikhs falcons," Tarre says. He started calling oil "the devil's excrement."

**Now oil is** Chávez's Exhibit A in his theater of balls and revenge. In early 2003 he fired the eighteen thousand striking PDVSA managers on TV, portraying them as traitors. The president enhanced the theatrics by blowing a whistle, shouting the name of a high-level manager, and roaring "Fired!" Really, he knows good TV.

I meet three of PDVSA's former managers at the Caracas Country Club, a magnificent building tucked into a private golf course in the middle of the city. It is so off-limits to ordinary people that the taxi driver spends ten minutes just finding the entrance. Passing the guard, I am in an elevated, all-but-colonial world, with obsequious waiters, yellow awnings, chintz couches, and painted tile floors. Yvonne's hardscrabble Caracas is a different continent. The older executive, Alfredo Gruber, wears a well-fitted gray suit, while the other two, Luis Pacheco and Armando Izquierdo, are dressed like suburban New England dads, in Woolrich jackets with Palm Pilots in the pockets. Drinking whiskey and eating fried cheese, Izquierdo tries to explain what it felt like to be fired.

"We were like a wild animal in a cage and Chávez was poking us with a stick. Ha ha ha! Having fun with us! He said we were doing wrong, we were an elite, traitors, and he kept bothering us with a stick until the wild animal started to react. That was us," says Izquierdo, acting out both the vengeful public with the stick and the cringing animal. "We thought what we did as oil workers was paying our debt to society. We discovered we were not popular and a good portion of the population had sticks against us." Telling the story, Izquierdo jumps out of his seat and jabs the air with an imaginary stick and then collapses, the hurt animal in the cage.

"Well," says Pacheco, the other Woolrich dad, "this had really been building since 1914."

All three men grew up in the oil company and were part of its peculiar culture. Along the shores of Lake Maracaibo, the foreign oil companies set up large camps, stratified by race and class, and proceeded to mold their workers into a hybrid Anglo-Venezuelan culture. Creole, which was the Venezuelan wing of Standard Oil of New Jersey (later Exxon), created a model of citizenship "that recognized the centrality of oil in Venezuela's development . . . and identified the progress of the nation with corporate interests," says Miguel Tinker Salas. Creole's American workers lived in walled neighborhoods called Hollywood, Victory, and Star Hill. Creole told its American workers' wives how to deal with their Venezuelan maids, and it told its Venezuelan workers' wives what kind of hats to wear to company socials. Creole's stores sold food at big discounts and encouraged workers to play team sports. Because the company feared a socialist uprising in Venezuela, it produced a magazine advocating democracy and culture.

When Venezuela nationalized the oil industry in 1976, the foreign companies left, but their Venezuelan workers stayed, and they kept their customs, their meritocracy, and their international orientation. "My personality was shaped by the industry," says Gruber, "task oriented, disciplined. It was an honor to work for PDVSA. I had a fun life doing my job." The other two were sent abroad for multiple degrees at MIT and Harvard.

"After 1975 there was a recipe for confrontation," says Pacheco, "because the revenues were not high enough to sustain the mirage of the state, and the nation thought it deserved more."

The government looked at the new oil company, which had pledged to be apolitical so as not to upset the pact between the two parties, and saw dollar signs. The government did not want to invest in the oil fields but wanted the money from them. When OPEC was unable to keep oil prices high between 1980 and 1988, the value of Venezuela's exports fell by more than half.

The country owed an increasing amount of foreign debts, and it was desperate for more money. But its oil fields needed huge investments just to continue producing, and even more to increase output and bring in more money. PDVSA couldn't get the money from the government, and having promised to keep the oil safe for Venezuelans, it could not ask foreign oil companies to invest. One manager took the prohibition on foreign investment almost personally. "The analogy was that you're an old nasty man with a lot of money. Because you don't trust anybody to take a check to the bank to buy you medicine, you just sit in your house, sick."

With nationalization, the old triangle of people, state, and foreign oil company was changed to include PDVSA. But was PDVSA part of the Venezuelan state or more like a foreign oil company? While the company was working for the state of Venezuela, the workers saw themselves as part of the world of multinational oil companies. To resolve the contradiction, PDVSA thought of itself as a first-world oil company in the service of Venezuela. "We could be considered a private enterprise," says Pacheco. "The goose that laid the golden egg for all of Venezuela. Nobody could touch us. So we started building our own empire."

The managers of PDVSA felt that the company could make more money by adding value to the oil. The company bought foreign refineries and invested in technology to make the country's heavy oil more profitable. In retrospect, these were decisions that should have been discussed by PDVSA managers, the government, and in public, but oil was the province of an elite. PDVSA managers were disdainful of government officials who arrived at their headquarters demanding to look at the books but without taxi fare to return to their own offices. "The industry has long-term strategies. It's a few years ahead of the state," said one manager interviewed by researcher César Baena. "The industry is the tail that wags the dog."

PDVSA's management complained that it was "a first-world oil company in a third-world country." In the early 1990s the company wanted to boost production so it could make more money even when oil prices were low. It found a loophole in the ban on foreign investment by declaring that some fields were "marginal" and put them out to bid in a program called *apertura* (opening). Desperate for money, the state acquiesced and only a few legislators complained, even though at least one of the contracts gave the state paltry royalties of only 1 percent. Behind the scenes, there were discussions about privatizing PDVSA and removing the state from the picture once and for all.

When Chávez was elected in 1999, he reversed the trend. He appointed the loudest opponent of apertura to be his oil minister—a man named Ali Rodríguez Araque, who had lived underground as a leftist guerrilla known as Commandante Fausto during the 1960s and '70s. Fausto (an appropriate name for a Venezuelan oil minister, if there ever was one) is a savvy diplomatic technocrat, and he set about getting the Saudis, Mexico, and Russia to cooperate in raising oil prices to around $20 a barrel. Then he became secretary-general of OPEC.

After the coup attempt in 2002, Chávez made Rodríguez Araque the head of PDVSA, but some in the company's management still didn't approve of the way the two were managing the company. In December 2003 management methodically planned a strike to stop production and close the refineries. With characteristic faith in their own abilities, they expected to reopen the facilities as soon as the president stepped down. "We were victims of our own self-importance," says Pacheco. "There was a leftover taste that the oil industry could topple the government." Instead, they were all fired. "The big lesson of these years," Pacheco continues, "is that if a corporation is not embedded with society, sooner or later things will happen."

Rodríguez Araque was left to deal with the strike and bring the oil company back up to speed without eighteen thousand managers. Ironically, he had spent much of his time as a guerrilla studying oil economics while designing explosives to blow up oil facilities. "How paradoxical!" says a Venezuelan friend. "To see Ali Rodríguez who was once this force with a dream of how to paralyze the oil company, and now to be paralyzed by the workers of PDVSA . . . Well, that's history in this country!"

"What you see through the twentieth century is this same screenplay," says Pacheco. "Over and over again it's the same screenplay. Like a Jungian myth."

For his part, Rodríguez Araque, whom I spoke with later, was unimpressed by the strikers. "They didn't have political experience," he said dismissively. "They didn't take into account the reforms in the army, the position of the population, or the other oil workers." He looked disgusted. "They overestimated their own powers."

Rodríguez Araque is an old-school nationalist, and he quickly instituted a new way to calculate the revenues that PDVSA owed the state. Eventually some of those apertura contracts were revised to yield the state 16.6 percent royalties and 50 percent income taxes. Later, in a speech, Rodríguez Araque announced that Venezuela and other oil-producing countries had to "overcome a colonialist past," saying that PDVSA's old management had acted in the interests of oil consumers (i.e., Americans), not Venezuelans. But PDVSA still needed foreign investment, and it was hard to walk the fine line between fervent nationalism and money.

Theater came to the rescue. On February 29, 2004, Chávez gave a speech in which he threatened to cut off the oil flowing to the United States. "If Mr. Bush gets the mad idea of trying to blockade Venezuela, or even worse, of invading Venezuela . . . the people of the United States should know that not a drop of oil would reach them from Venezuela." (It was the same speech in which he called President Bush a *pendejo*, which translates variously as "idiot," "wimp," "jerk," or "pubic hair.")

Chávez has an endearingly crackpot way with Yankee baiting, and his remarks played well. Initially, it was hard to tell whether he really feared invasion or whether it was a good way to energize his supporters. His petroposturing seemed ridiculous, but he liked to give the impression he was just crazy enough to do it.

But in a way it didn't matter. His threat to cut off the oil had in fact already been carried out, and was being carried out, by the slow eclipse of Venezuela's oil production by politics. In 1970 the country produced 3.8 million barrels of oil a day; today that number is somewhere closer to 2.5 million and probably even less. The problem has little to do with the oil itself; Chávez is sitting on 77 billion barrels of reserves, the largest outside the Middle East. The issue is the dicey nationalist politics of investing money in the oil fields. Chávez either has to find a way to invest more money in Venezuela's wells, or he has to be content with pumping less oil.

As it turned out, the motives for the president's "pendejo" speech were more complicated than they appeared. Less than a week later Chávez and PDVSA's Rodríguez Araque welcomed Chevron Texaco executives to the presidential palace to sign an agreement for offshore gas exploration. Though the new revenue scheme required that the American company pay a high percentage of profits to the government, rumors were going around that the country, desperate for investment, had secretly accepted much less than it claimed. Critics on the right scoffed. "It's schizophrenic," says Francisco Monaldi. "You have the leading advocates against apertura offering oil deals to foreign companies at the worst possible time. The logic is to hold power."

"**An opera,** a three-act tragedy in ten days" is how Douglas Bravo describes Chavez's speech and the agreement with Chevron Texaco. The "pendejo" speech and the threat, he says, were a populist distraction for the masses while the real contracts were signed behind the scenes.

Bravo is an unlikely Chávez critic because he is one of the country's legendary leftist guerrillas and was in fact Ali Rodríguez Araque's commander for the many years they spent in hiding. Now Bravo lives openly with a crowd of family and supporters in a raggedy high-rise in Caracas. An amorphous crowd treats the apartment as if it were still a hideout, and Bravo, an energetic fireplug in his sixties, bellows happily into his cell phone walkie-talkie style. I meet Bravo early one morning and he offers me a frosty beer, wrapping it delicately in a napkin, before settling in to discuss his beef with the president. Nationalism, he says, legitimizes Chávez's government, but his actions are just the opposite. "These oil policies are antinationalist and antipatriotic," he says. "If another government had signed these policies, Chávez himself would have planned an insurrection." The very thought makes him chortle.

"The irony of destiny is large and life makes tricks on people," Bravo says, predicting that mismanagement of PDVSA's fields and the actions of his old friend Fausto will ultimately cripple the oil company so fully that the national patrimony will end up in the hands of the Americans. Still, he says, he supports Chávez against the Americans, and he assumes that the next few decades will be a slowly heating struggle over control of the country's oil resources. He offers me another beer, saying jovially "It's going to be a long war."

**Lake Maracaibo,** the source of Venezuela's wealth, is a great gray plate, forty miles long, home to more than 37,000 miles of underwater pipelines dating back to the 1920s. Maracaibo is not a lake in the conventional sense—with kids in bathing suits, beach balls, and pleasure boats. Better to use the word "facility" to describe its bristle of electrical lines, oil platforms, strangely human silhouettes of pipe, and otherworldly outlines of collection stations gleaming through the mist. There is something mysterious about the perspective on the lake. The installations look like bathtub toys as you approach until all of a sudden, without any transition, they loom overhead.

I'm in a peapod of a fishing boat with Jorge Hinestroza, a professor at the University of Zulia. Jorge studies the ecological history of the lake, and he is earnestly fretting over the proliferation of obnoxious water lentil when we are joined by an enormous seagoing coast guard cutter with men on the bow carrying machine guns and shouting "Are you the professor? We're here to protect you!" Our boat bobs in the cutter's wake. Jorge is a compact man with gobs of curly hair and a love of the comic double take, the consummately bemused straight man. The arrival of the coast guard overrides all of his comic asides. He looks distressed and confused. "I'm embarrassed," he tells me. "I told a friend who's a manager at PDVSA we'd be on the lake this morning. And this is the result." With every flick of our peapod, the great cutter flicks too, like a shark mimicking the movements of a minnow.

At the time, I thought this oversize escort was no weirder than anything else on Lake Maracaibo. The escort is no more bizarre than the fact that oil extraction is causing the land around the lake to sink by eight centimeters per year, for example, so that now the lake is contained behind a tall dyke with a whole town of 35,000 huddled against it, ready to drown if the dyke breaks. And it is no more incongruous than PDVSA's village of Southern Gothic homes on the east side of the lake named Hollywood. And it wasn't any crazier than the fact that two groups of job seekers are having a gunfight that morning in front of PDVSA's main office and the taxis are refusing to go there. So I didn't give the boat much thought.

Over the next few days Jorge gives me a tour around the lake. We visit the flare near where he grew up. "If I couldn't sleep I'd go to the flares in the middle of the night. They're a little hypnotic." He and his friends swam in the lake until the late 1950s, when the water became so polluted they had to clean themselves with kerosene afterward. He talks about the dangerous pollution patterns around the lake. "This state is an official sacrifice zone for the revolution," he says with a chuckle.

Jorge won't let go of the coast guard boat. He keeps chewing on it, trying to figure out what was wrong with it. At first he says the boat was a waste of resources. He wonders if his friend suspected he'd sabotage the oil facilities. But that is ridiculous, because everyone knows Jorge is with the struggle if anyone is. And later he wonders if it was a show of "the people's revolution" and the ship had been dispatched to "protect" us and show us the great benevolent power of the "pretty revolution." Then he wonders if maybe the manager was trying to show that he was a *chaviverde* (green Chávez-supporter) as Jorge was a *comiflores* (flower eater). Still later he wonders if it was some kind of purchased populism—Look, little guy! You're part of history! He thought that was very far from the ideals of the Zapatista revolution. Later he e-mailed me, "To see a 'big power' protecting a 'powerless thing' will be always weird."

I enjoyed watching the twists and turns of Jorge's take on the ship because his thoughts were more revealing than that secretive lake. I carried away two points. 1: The petrostate's swollen emphasis on politics turns even the smallest aspects of everyday life Machiavellian. Every boat ride must be scrutinized like a pool table with a ball in front of every pocket: What is the purpose? What game is in play? How should I protect myself and my family? And 2: What kind of people's revolution starts with a $40 billion oil company?

**By noon,** the gunfight has died down, and I go to PDVSA's main office on the lake. On the wall behind the receptionist's desk hangs a painting of the big gusher of 1921, which led to the creation of the petrostate and the situation we are in now. Done in a folkloric style, the painting shows local farmers rushing an effigy of Saint Benito toward the spewing well in the hopes of removing its greasy curse from their croplands. It's an odd choice of decoration for an oil company lobby, suggesting that PDVSA is trying to do the best it can with the devil's excrement.

The lobby has security and a metal detector, a new addition since the strike. In and out of the building people greet each other by talking about *La Lucha,* which they translate as both "the struggle" and "the process"— a standard greeting for Chavistas, though usually delivered with fond and comic exasperation.

A young woman walks toward me dressed in low-slung black jeans held up by a belt decorated with a large black leather rose. "We're in the Matrix," she says in greeting, referring to the sci-fi movie starring Keanu Reeves. Fabiola, my assigned representative from PDVSA's PR department, speaks in short bursts, as if gulping for air. In the year since the strike she's been working eighteen hours a day with hardly any time off. "PDVSA is like Hollywood for PR people," she continues. "Disney." She takes a call from her seamstress, who is angry because she has not come to have her wedding dress fitted and the ceremony is only weeks away.

"My stomach is. KILLING. Me. But. That's usual," she says. She decides we need to go to lunch, complains she can't have coffee or smoke because of her ulcers, but orders a coffee. When the strike started, she says, she volunteered to work for PDVSA for free for two months. Now she has an official job, and a pending marriage, and she's fortunate to be stressed out by work because everyone else in her well-off social set is unemployed, getting divorced or depressed or worse. "Two thousand three was the year for lawyers, psychologists, and cardiologists," she says, smoking.

The problem is that in her circle, she is one of the few who support La Lucha. There have been repercussions. She mimes talking on a cell phone with her thumb and pinky spread. "All right. I will. Never talk to you again!" She slams the imaginary phone down.

We head across the lake to the city of Maracaibo to meet PDVSA's regional director. Javier Alvarado is tall and statesmanlike. He looks slightly baffled by the five other people who enter his office for my interview, apparently sent by different branches of the PR office.

In the past, he begins, the managers at PDVSA kept the customs of the foreign companies their parents had worked in. "You can see the homes behind the fence—for us," he says, waving across the lake toward Hollywood. "We gave the communities only what was necessary to do our business. And we felt comfortable saying that our business was to produce oil at the minimum cost. We recognize now that we must go beyond that fence in the way we do things."

The company has stopped buying equipment overseas, he says, and has started trying to develop industries at home. "We used to think, Buy the best! But now we think about what we can do to develop the country. It's a switch in your mind," he says. Another "switch" is that PDVSA has stopped comparing itself to the multinationals and settled down to being a national oil company. "We've realized our stakeholders are the 25 million people of Venezuela," he says, "so we must be in complete alliance with the Venezuelan state." The plan, he says, is to increase Venezuela's oil production to 5.2 million barrels a day. New oil investment projects are judged not just by their financial returns but by their contribution to GDP and long-term job growth as well as guarantees of water and electricity to nearby communities.

I ask him what changed his mind. How did he join La Lucha? What made him less interested in competing with other countries and more interested in life outside of PDVSA's fences? He deflects my question each time I ask it.

He wants to talk about the company's stake in providing Chávez's so-cial programs. "It's not really our role," says Alvarado, "but if we want to have better numbers within PDVSA, we need to improve the quality of local education." Fair enough, though an unlikely task for an oil com-pany. But he continues, "We need to drag the country behind us." The normal path would be to give the oil money to the ministry of education, he says, "but most of the money will get lost along the way. What's the fastest way to get that money to the community? You rely on the educa-tion minister to do their thing? Or the way PDVSA does it? We feel we are more efficient."

It's a strange speech, implying that the oil company will eventually take over the responsibilities of the incompetent state. PDVSA is going be-yond the fence and extending the fence line to include all Venezuela. The latest change in the petrostate triangle is to merge the state and the oil company into one behemoth, making the state structurally even more de-pendent on oil. It does seem ironic that a group of Marxists from the hills would have reinvented the petrostate as a benevolent petrocorporation.

Critics worry that combining the state and the oil company may jeopardize the oil money. "When you look at the structure of the cash flow and the social spending, you see reasons to worry about the health of the oil industry," says an American oil consultant in Venezuela. He says that in addition to the official social programs, social aspects are being added to other contracts. The developer of a gas field, for example, has to build a power plant for use by the local community. "The political people in PDVSA don't care if the company is profitable," he says, "they just want financing for their revolution." The company announced that by the end of 2004, it had spent $4.35 billion on social spending and nearly $3 billion on field investments.

Later, as I stand on the dyke that holds back the waters of Lake Maracaibo, it's possible to imagine a future where the internal politics of petroleum have eclipsed the current logic of the oil market and Venezuela's oil exports to the United States trickle off. Ever since the very first days of the oil age, in the 1860s, periods of high oil prices have been followed by investment binges, which put more oil on the market and lowered prices. But in Venezuela, the state and the oil company are in competition for the same scarce resources, and as long as political in-stability around the world keeps oil prices high, there is no pressing need to invest. By 2006 there were even rumors that PDVSA's oil production had fallen off dramatically, and the company was buying oil from Russia to fulfill its long-term contracts.

Most forecasts call for a huge increase in OPEC oil production in the next twenty years. One estimate says the world needs to invest $16 trillion in energy supplies by 2030 to keep up with demand. These projections don't take into account the fact that OPEC's production has fallen since 1978. Like Venezuela, many petrostates have been reluctant to invest money in oil production or to accept foreign funding. Kuwait, for example, has planned to let international oil companies invest in increasing its oil production since 1997, but the details of the deal have gotten bogged down in a parliament fearful of foreign intervention.

High prices have made the oil even more useful politically. In 2005 Ali Rodríguez Araque moved from head of PDVSA to become Venezuela's foreign minister, engaging in a constant roster of petrodiplomacy. What started as trading oil for Cuba's doctors and then for Argentina's beef became much more ambitious. Later that year, he told me that oil is an economic and social force in the country, but it also has a political component. "The link between sovereignty and oil is very strong. In Venezuela, both Gómez and Pérez Jiménez were supported by the foreign oil companies. I hope the leaders of the U.S. realize that Venezuela is a foreign country. We have legitimate rights. We need this market and the States need the supply from Venezuela."

By 2006 Venezuela expanded its political reach, even covering $2.5 billion of Argentina's debt. The country was selling oil at a discount everywhere from New York's Bronx (to poor residents hurt by high oil prices), to the Caribbean and Uruguay. Chávez was creating common interests among Venezuela, Argentina, Bolivia, Cuba, Nicaragua, Ecuador, Chile, and Brazil with trades of oil, pipelines, and influence. These countries were further united by a round of neoleftist thought, partly inspired by Chávez, partly as a reaction to failed neoliberal policies. To unite Latin America, Chávez sponsored TeleSur, a TV channel that aims to be a Spanish-language Al Jazeera. Farther abroad, the country is using oil to cement deals with China and Iran. Chávez was quoted as saying that Venezuela "has a strong card to play on the geopolitical stage. It is a card that we are going to play with toughness against the toughest country in the world, the United States."

Chávez may have been playing cards, but the United States prefers to think that the game is still as it always had been—a simple exchange between a big country with a lot of money and a small country with a lot of oil. It is the "limited rationality" of the gasoline consumer writ large: In the United States, the cost of oil is confined to the numbers on the pump, not to intangible politics. And as petrostate politics prove more and more pricey—in terms of both the cost of crude and the insecurity of supply, the United States has fewer options.

Increasingly, oil is the only connection between the United States and Venezuela, a fact that, paradoxically, makes the oil flow more vulnerable. In late 2005 Exxon, which as Creole had been the architect of the Venezuelan state in the 1920s, sold its stake in an oil field rather than agree to changes that Chávez demanded. In 2006 the company was kicked out of a Venezuelan petrochemical project and a newer oil project was stalled. As Douglas Bravo said, "The irony of destiny is large."

After I left Venezuela, it came out that the United States has actually given considerable thought to invading. The Department of Defense labeled Venezuela a "rogue" nation in 2004 and drew up plans for military action should the country pose a "pop-up" threat. According to *Washington Post* military analyst Bill Arkin, Venezuela's closeness to the United States, and its oil, and the "leftist" nature of Chávez, gave the plans the added strategic priority of being related to homeland security. At the moment, it's hard to imagine that the U.S. public would support an invasion of Venezuela, but what is striking about the plans is how well they serve Chávez. In early 2006 Secretary of Defense Donald Rumsfeld echoed the Venezuelan opposition by comparing Chávez to Hitler. For all its hyperbole, sinister tone, and likely benefit to his opponent, Rumsfeld could have been reading a script written by Chávez himself.

**Jorge picks** me up in his massive green Impala. Door handles stick out like pieces of broken bone. All of Maracaibo drives these huge rattletrap cars, burping black smoke from gasoline that costs 9 cents a gallon. Nothing points out the tensions in the oil state better than cheap gasoline: What's good for the state is *not* good for the nation. Cheap gasoline encourages pollution, waste, and smuggling over the border to Colombia, and it prevents the country from selling the oil abroad for more money.

Citizens celebrate their patriotism by plundering the patrimony. It is a system constantly at odds with itself, La Lucha at its worst.

Many miles from Maracaibo, Jorge pulls off the road into the scrub toward a petrochemical plant named El Tablazo. The Impala whines into the sand, jouncing around the hill past the NO ENTRY sign until we are facing the ruins of a small town studded among tall brush.

A flare roars overhead, burning the gases from the plant and filling the scrub with an enormous hiss. I have to lean close to Jorge so I can hear what he's saying. "When this was the village of Hornitos, they had two flares going," he yells. "Everyone here spoke by shouting. They didn't have night anymore because they always had this huge candle." He points to the flare. "The rain became acid rain and the whole community would lose their clothes if they hung them outside to dry. Two to three times a month a release of chlorine gas would send a thousand people running from the village looking for air." Jorge was describing a perfect hell, a perversion of life so complete it sounded like the plot from a novel by Gabriel García Márquez. It was hard for me to imagine that these random walls had once been a village. The town smelled of sagebrush and faintly of cleaning products—as if someone nearby were scrubbing tile. "Even now," he yells, "I can't understand why they stayed here."

Hornitos was originally a fishing village on the edge of the lake. The petrochemical plant was started in the 1950s and continued in a mess of corruption until it was complete in the 1970s, costing billions of dollars. Jorge says the villagers accepted what they got. And as people began to get sick from chemical releases, they didn't complain. Even when they had spontaneous abortions. Even when eight women aborted in their eighth month. Jorge's students began surveying the village in the late 1970s. Eventually they discovered that people had hidden children with birth defects in their homes.

"It was so sad that I cried," Jorge said. For years he tried to puzzle out why the villagers stayed silent. "I wondered for a long time why they didn't realize the trend. I don't think it was lack of education, or lack of understanding cause and effect, or shame. I think they didn't want to notice because they were powerless to move." The external locus of control again.

Now he wonders if Venezuelans aren't somehow all in the village, trapped under the big specter of oil. "PDVSA is the main patron of this country, and our whole modern history is oil," he says as we drive back toward Maracaibo. "PDVSA rules Venezuela." In his view, the state has always seen itself as the ideological appendage of the oil company—smoothing things over, making sure that the one true business of the country continues at any cost. "PDVSA is the power. Whatever is economic, strategic, or political it has to do."

The old PDVSA and the new are all too similar. When the villagers of Hornitos protested in 1992, the military threatened to try them for treason because they were endangering the economy and national security. Jorge briefly went to jail. Though Chávez's government has done a better job of distributing the oil wealth to the poor, it has not abandoned the idea of traitors. In 2004 government officials accused environmentalists of being a "green mafia" supported by international oil companies and the CIA. An indigenous activist against coal mines has been called a terrorist. Around the lake, newly empowered Bolivarians invoke the constitution and the Bible but they cannot question the primacy of oil. And so, La Lucha goes on and on.

**I'm back** in Fabiola's chauffeur-driven car, and we're whizzing across the lake again, passing through the scrabble of ragged towns along its edges. Fabiola, who is redoing her makeup, says she noticed that Alvarado didn't answer my questions about why he is with La Lucha. She says she asks herself the same question and she only knows that she believes. "For all of us it was something that made sense. The old way didn't work."

The old way didn't work: Venezuela's triangle seems broken beyond repair. It's a measure of how bad things are that La Lucha, the great planless struggle, looks attractive. People can't seem to imagine a country that isn't dependent on oil, and the contradictions inherent in the oil economy necessitate a struggle. Where it is all going hardly matters, a majority of people have given up on the old way. This, I think, is the appeal of La Lucha.

And La Lucha isn't confined to Chávez supporters. I'd met some of the eighteen thousand fired employees, who call themselves the people of petroleum. Now unemployed, they were at an upscale coffee shop in Maracaibo, planning what they called their comeback. The former managers, unfamiliar with the joblessness that affects about a third of

Venezuela, got up every morning, put on pressed shirts, picked up their briefcases and Mont Blanc pens, and headed to a coffee shop. One, educated in Texas, used the word "purdy" a lot. As in: "We have a purdy flexible constitution. It says that when the people are not getting democracy, they have the right to fight." I'm incredulous, staring at the collection of electronic Mercedes keys, cell phones, and designer pens on the café table. You're planning to fight a war over the oil company? I cannot imagine them sneaking around in the woods in their pressed shirts. Yes, he says, and when they win, they have a plan to make PDVSA better than ever. As soon as we're in the car, his friend leans toward me and says, "Actually, it's been more than a year. I have a family. If Chávez stays in power in the referendum, I'm thinking of moving to Houston."

Fabiola says everyone is in the Matrix, which seems to be her description of the strange hybrid of oil company, collective farm, encounter group, and political animal that PDVSA is becoming. When she worked at the personnel department, she spent Sundays watching Chávez's interminable show, *¡Alo Presidente!*, to see which way the political winds were blowing and figure out the correct answers from job applicants. Her family yelled at her because the show is obnoxious. She mimes slamming a door and then herself, arms crossed, determinedly watching the TV.

She seems less fazed by the specter of Chávez's personality cult than by the emotional toll of the revolution. Meetings at her office are horrible, she says, because people scream at each other. She asked her boss about it. "He said it was. Catharsis. We're in the Matrix." Her mother needed a leg operation, so Fabiola put her on her insurance. But then her boss noticed her mother's name on a list of people who'd signed the petition to recall the president. He brought this up in a meeting and tried to pressure her. Fabiola says she fought back. Still: La Lucha, La Lucha. She's with the process. Everyone is doomed to stay with the struggle until oil finally makes good on its fantastic promises.

An older worker leads us on a giddy tour of the marina, showing the pylons the company now makes rather than buying from abroad. He shows us the red hard hat Chávez wore when he visited, and the machine shops where the workers' co-ops build desks at night. And then there is the first boat they made, named the *Bolivariana*. "Everything's *Bolivarian* here," he says. "Bolívar said the best government is capable of giving the most happiness."

# 8   CHAD *BETTING ON THE LION PEOPLE*

Chad sits in the middle of Africa, a landlocked chunk of desert and sahel. Flying in from the north, we cross over the tiny electric lights of the last villages of Nigeria and Cameroon before the plane begins descending into Chad's velvety darkness, as if we've glided off the edge of the world. N'Djamena, the capital, is home to 700,000 people, and the city's generators are chronically down.

On the plane, the oil workers from Texas and Oklahoma shift in their seats. Even for oil guys, Chad is pretty far off the map, but it became the world's newest oil exporter a month before my arrival in November 2003. Chad is on the frontiers of oil—the risky places where multinationals now hunt for reserves and profits because the easier reservoirs have already been developed.

N'Djamena's airport is the size of a small-town post office, and a generator keeps it cheerily lighted. The luggage pile (there is no belt) is surrounded by a tall crowd of Chadians in business suits, murmuring to each other in French. If they all seem on familiar terms, it's probably because only a very tiny fraction of people can afford to fly in a country of 9 million people with a per capita income of $170.

Out of their seats, the oil workers' barrel chests taper to their wispy cowboy boots, so they look like anxious ghosts. Exxon has invested $3.7 billion in Chad's oil project, building an oil field and a 660-mile pipeline from southern Chad to the Cameroonian port of Krebe—the largest investment in Africa ever. And it brought together Exxon and the World Bank in a scheme to make oil money develop Chad, one of the poorest countries on earth.

Just a month before my arrival, in October 2003, Exxon, the World Bank, and Chad's president, Idriss Déby, proclaimed the beginning of a new era, a model for the world. This is to be the country that escapes the petrostate curse that has plagued every developing oil country since Venezuela started exporting oil in the 1920s, but if anyone thinks it's ironic that Exxon is involved in this attempt too, they don't mention it. Exxon is cautious, and oil industry publications note that the company has placed two sets of controls on that pipeline, one in Chad and the other in Houston or Paris.

I step out of the airport into warm darkness that smells of sun-baked sewage, charcoal smoke, and not unpleasantly, people—a mixture of faint but distinct musks occasionally interrupted by a trail of perfume. Kerosene lamps make circles of light around the peddlers on the steps, and the taxis' headlights illuminate long cones of swirling yellow dust above the unpaved street.

I am traveling with a French photographer, on assignment for the Italian magazine *Colors*. We get into a shuddering Peugeot taxi. The car's cockeyed headlights graze walls topped with jagged glass and metal spikes, stores barricaded with iron doors, trees and more walls, and the occasional compound lighted by its own generator. On a warm night in the middle of Ramadan, when I'd expect people to be hanging out eating voraciously after a day of fasting, the streets are strangely empty.

We reach a ratty cinder-block hotel with its own generator. Once we're inside the courtyard, the clerk says with a wide smile and gentle diction that we cannot leave the hotel because we will most certainly be attacked by bandits or corrupt police, possibly one person who is both, or possibly one and then the other. We stand in the courtyard for a while, under the electric light, slapping mosquitoes with the watchmen. Finally I go to my room and dutifully slide shut the four dead bolts.


**When the sun** rises, a banner reading L'ERE DU PETROL hangs in the middle of the dusty yellow roundabout next to the hotel. "The era of petroleum." The expression is matter-of-fact, neither celebratory nor foreboding, simply an announcement: No turning back now, it's petroleum time. Start the music. Driving around the banner are white SUVs belonging to oil companies and nongovernmental organizations, battered vans with people hanging off their edges, bureaucrats on bicycles. In the outer reaches of the orbit, a stream of women carry small pink bags of sugar on their heads.

The walls behind the roundabout are bullet-pocked. Most of the older buildings in N'Djamena are scarred by nearly thirty years of war that followed Chad's independence from France in 1961. The first president has been described as Ubu-esque by historian Sam C. Nolutshungu in one of the few modern histories of the country, tellingly titled *The Limits of Anarchy.*

Nolutshingu calls Chad one of the world's "grey areas" or "strange places" where the state is caught between emerging and disintegrating. Long described as a vacuum by African diplomats, Chad has spent most of its history muddling along in a violent haze with intervention from France, Libya, Sudan, and the United States. During the 1980s, Libya invaded Chad, and a collection of warlords fought back with more than $50 million in military support from the United States. Déby, the current president, is a former warlord who installed himself after a coup in 1990, then held an election in 1996 to legitimize his position.

With only three hundred miles of paved roads and limited telephone service across land three times the size of California, Chad is understudied and unknown. Statistically, it is the tenth poorest country in the world, and 80 percent of the population lives on less than a dollar a day, the official marker for poverty. The conventional poverty scale may not be sensitive enough to register just how poor this place is, because I meet several people who are living on less than 25 cents a day, and they appear to be no worse off than their neighbors. Chad doesn't even have much of a trader class, and in the scraggly market, it takes me ten minutes to change a Chadian bill worth less than $5.

The constriction and poverty of Chad's economy make everyday scenes here surreal. At N'Djamena's border, guards with sticks shake down everyone who crosses the bridge from Cameroon, demanding bribes of "lunch money" from people carrying lemons and soda. But a steady stream of determined people on three-wheeled wheelchairs cross the bridge in both directions. Though all of them are missing limbs or disabled, they are surprisingly well dressed. One man whizzes past wearing radiant lime-green robes, and a woman clanks by in an impressive headdress. The French-designed three-wheeled chairs are propelled by small cranks on the front, which the chair owners pump rapidly by hand, with a determined set to their jaws, leaving a spurt of dust behind them. The handicapped are allowed to smuggle two bags of sugar daily from

Cameroon (where it is cheap) to Chad (where it is expensive) by presidential order. This is described as a "social program," and in fact it leaves the wheelchair riders fairly well off, as their profits are $1 a day. The head of their union has a cell phone, but he is too busy to talk during Ramadan.

In the French-designed center of the city, two traffic lights stand useless, symptoms of a country that seems to be undeveloping and moving backward. The generators that have powered the city since the 1950s rarely work, victims of corruption and lack of fuel. Nearby stand the modernist ruins of a gas station, chrome gleaming dully under dust. Now gasoline comes from Nigeria's black market. Peddlers fill 750 ml Pastis bottles with green diesel and pink gasoline and arrange them on rickety tables by the side of the road. The bottles glow greasily in the heat, with all the allure of a poisonous aperitif in a fairy tale.

In the market, women wear dresses made of fabric boasting TCHAD, NOTRE ECONOMY EST PETROLEUM, but this is wishful thinking, as petroleum hardly exists here. According to UN figures, Chad has the world's lowest per capita fossil fuel consumption—96 percent of its energy comes from wood and dung. Soon this unelectrified country will be exporting 250,000 barrels of oil a day. That's roughly the amount of oil used daily by Austria, which has a similar-size population and has used oil to fuel a marvelous infrastructure and lives of extraordinary ease, organization, and predictability for its citizens. Despite all the hope placed in it, Chad's oil project, in contrast, is expected to bring in just $80 million a year for the first four years—less than $10 per person.

At night, kids gather outside the oil companies' security lights to do their homework. A junior in high school, Mahamat is sitting on the anti–truck bomb barrier outside of TCCP, Chad's oil services company. Barbed wire and greenish fluorescent tubes arc over his head. He shows me what he's studying: computer science. With the smooth face of a child and the purposefulness of someone much older, Mahamat tolerates my questions with a thin-lipped smile. He won't admit to the futility of studying computer science in a city without electricity. He hopes to work for the government, he says. He's studied at the barrier six days a week for the past two years, and he has great hopes for the oil project. "I've heard about the petrol," he said, "and it's going to change a lot of things. It'll be the development of Chad."

He uses the word "development" in the magical sense, the way most people in Chad use it—meaning everything to everyone and nothing in particular. Development could be cream to heal ringworm, a village well, a network of roads, trustworthy policemen, a functioning economy, or a state like Sweden: It's a catch-all. Mahamat said he wanted electricity, better schools, and security from ethnic violence and discrimination. I asked him what petrol looked like. "A benediction." He grinned for the first time. "A blessing."

**Chad's poverty** is exceptional, but in the scheme of World Bank investments in oil projects, it's not unusual. Between 1992 and 2004, 82 percent of World Bank investments in oil in developing countries were for export of that oil to developed countries. This has been a deliberate strategy on the part of the United States, which has more influence than any other country in the Bank, to advance American energy concerns. In 1981, after the second oil crisis, the U.S. Treasury Department urged the Bank to use its "neutral stance" as a "development advisor" to expand into oil projects to "enhance security of [energy] supplies and reduce OPEC power over oil prices." In effect, the Bank became an instrument of U.S. policy, just as the oil companies had been in Venezuela sixty years before.

For more than a decade, this approach put a lot of non-OPEC oil on the market, which kept prices down, but by the late 1990s, it was running out of steam. By 2005 Exxon's actual production was less than it was in 1998. Though non-OPEC oil was still rising by more than a million extra barrels a day every year, about 800,000 of those barrels were from Russia. The yearly rise in non-Russian, non-OPEC oil was around 200,000 barrels a day—something along the lines of Chad.

Africa, which has more undiscovered oil than anywhere else, has become a key part of U.S. strategy. As the curiously named AOPIG—African Oil Policy Interest Group—puts it, "African oil is not an end but a means to both greater US energy security and more rapid African economic development." The United States now gets more oil from West Africa than from Saudi Arabia, and the hope is to get a quarter of U.S. imports here by 2015. Energy is almost the only way U.S. companies invest in Africa—two-thirds to three-quarters of American investment in Africa is in energy.

As the world's gray areas and strange places become more the norm than the exception for oil development, there is pressure to stabilize them so that they too can begin pumping oil northward or westward. If the Chad model works, people talk of applying something similar to São Tomé and Principe, Iraq, Angola, Nigeria, and Kazakhstan, and onward through the Central Asian and Caspian states. Mahamat is waiting for a "benediction," but other people are waiting for the oil.

**Oil was** discovered in Chad in the 1960s, but it was a relatively small quantity, of poor and sticky quality, and far from either markets or a port. Quite soon it was located in a war zone.

"Yes. In 1971 we got the first barrel of oil out of the ground and we had a party. We thought it would change things," says Abdoulaye Djonouma. Djonouma is a prominent trader in gum arabic and president of Chad's Chamber of Commerce. I visit him in an office connected to his warehouse, while two gazelles munch on grass in the courtyard. His business card is roughly the size and texture of a wedding invitation, and it says he is also a Commander of the French Legion of Honor and Chad's first registered pilot.

Djonouma is tall and imperial, standing with his chest outthrust, and at seventy he retains the air of the young man who was the hope of independent Chad. But the years since the bright promise of independence and oil have given him the soul of a pessimist. Oil, he says, brought about economic and agricultural collapse in Nigeria and Gabon. For Chad, which has fewer resources, he fears worse: militarization. He ticks off all the former French colonies that have become militarized. Virtually all. (One study found that oil-exporting countries spend between two and ten times more on their militaries than other developing countries.)

"The African countries started well in 1959," he says, "but then it was coup d'état and coup d'état and dictatorship." In 1963 he traveled to Washington and met JFK. He'd show me the photo, but it was stolen or destroyed during civil war in the late 1970s.

In 1994 Exxon (known as Esso in Chad) realized that there were a billion barrels of oil under the country's mango trees. A billion barrels is considered the "magic number" for profitability. In the late 1990s, the company negotiated another magic number—a contract with the government giving Chad 12.5 percent of the royalties from the oil. An Exxon executive told a reporter for the *Wall Street Journal* that the agreement was reached in a marathon three-week negotiating session of twelve-hour days at an Exxon facility outside Paris. A more established oil state would have hired an outside negotiating team, and probably would have gotten a better deal. The details of the contract are not public, but two unusual aspects of it are that Exxon is permitted to import equipment to Chad without paying customs and the company avoids income taxes during the years of peak production. PFC Energy, an oil and gas consulting company, estimates that Chad's take is 28 percent of the oil's value; in contrast, Angola's take from its oil is 60 percent and Nigeria's is 80 percent.

Exxon set about drawing up plans to build an oil installation and a pipeline from Chad to the seaport of Krebbi. The total cost would be $3.7 billion. Shell and Elf joined Exxon as partners in the project, and then they went looking for funding.

Commercial banks reviewed Exxon's magic numbers and then looked at Chad's political situation and saw a risky business deal. Too risky to risk. So Exxon asked the World Bank to join the project, to lessen its political risk (because, in theory, the World Bank could hold back other funding that the poor country needed) and to act as a "catalyst" for other funders. "Chad could not afford to lose its World Bank contacts and we could not afford to go it alone," an Exxon executive explained to the *Wall Street Journal*.

When the World Bank ran the numbers, it too saw a risky proposition. Its report gave Chad a "Significant" rating, which meant its chances of failure were "less than 50 percent probability but significant adverse impact"—a vote of confidence just slightly better than a coin flip.

If you are looking to bet on an oil project in a developing country, the odds are clearly against. At Stanford, Terry Lynn Karl's analysis of Venezuela's economy during the 1970s and '80s shows that countries whose economy is dominated by oil exports tend to experience shrinking standards of living—something that Chad can hardly afford. Oil has opportunity costs: A study by Jeffrey Sachs and Andres Warner showed that of ninety-seven developing countries, those without oil grew four times as much as those with oil. At UCLA, Michael L. Ross did regression studies showing that governments that export oil tend to become less democratic over time. At Oxford, Paul Collier's regression studies show that oil- and mineral-exporting countries have a 23 percent likelihood of civil war within five years, compared to less than 1 percent for nondependent countries.

Aidwise, Chad was a basket case, with over $1 billion in debts by the late 1990s. For every dollar the country got from aid in 1998, it had to spend 43 cents servicing its debt. Oil, some in the Bank thought, might be a way to get Chad out of the debt spiral, and if it was managed right, perhaps the project could deliver more for the people of Chad. The one thing oil does is help poor countries pay back their debts. That's been a successful strategy for the International Finance Corporation, one of the World Bank's lenders, which has made the most profit on its investments in oil and gas projects.

One of the people who campaigned for the project was Donald Nor-land, a former U.S. ambassador to Chad. When I spoke with him in early 2005, Norland talked about his dedication to the project. "I got deeply involved with the legislation and the World Bank. For five years I was very attached to this. Oil is not the problem—any country is perfectly ca-pable of squandering development money. I get upset when people say that there's something tainted about oil. I saw Chad as a lab experiment, as a complex public-private partnership to see whether or not it could work. If it doesn't we're in trouble."

The Bank's lab experiment in Chad tried to redesign the petrostate triangle by inserting new players. Between the state and the oil money, it wanted to put a panel of citizens (called the "college") who'd decide where the money went. Then the Bank would stay involved as an exter-nal auditor to advocate for the rights of the poor and enforce first-world environmental standards on the project.

The Bank made an agreement with Chad's president to write a new oil law requiring that oil royalties (which make up about a little more than half of the income from the oil; the rest is in taxes) go directly to a bank account in London. From there, Chad's debt payments would go di-rectly to the World Bank; 10 percent would go to a savings account for "future generations," and 15 percent would go directly to the govern-ment. The remaining 75 percent would go to the "college," which was expected to spend the money on the priority sectors of health, education, social services, and rural development. The agreement was sensational in that Chad's government was giving up part of its sovereignty. But Chad's government wanted the project, and so it signed on.

Before the World Bank could go ahead with the plan, it needed the agreement of the United States. World Bank loans must be reviewed by branches of the U.S. government and then approved by Congress. USAID (United States Agency for International Development) looked at the program and wasn't keen on it. Consultant John Fitzgerald wrote that "governance was weakening, civil conflict and risk of famine were increasing," among other issues, but his comments were excised by the Treasury Department, leaving a description of "a model example of a project." "It's a model," Fitzgerald says, "but so was the *Titanic*. It was a great, modern design, but the details brought it down."

The Chad Cameroon Development Project, as the oil project was known, had failed to impress either as a business proposition or as a development scheme. Exxon's partners Shell and Elf backed out, without saying why. Exxon brought in Chevron Texaco and Petronas, the Malaysian national oil company.

And then there was Déby. When the new Exxon consortium paid him an advance of $25 million, he immediately spent $4.5 million of it on weapons to fight rebels within his own borders. The president of the World Bank personally called Déby, who dutifully gave $13.5 million in unspent funds to the college to distribute. Some viewed this as a success, because it showed that the Bank's moral suasion could keep the president on the right path.

When the project came before a congressional committee, a long line of academics and environmentalists said that the project was a bad idea. Former ambassador Norland took a different tack by changing the premise of the bet. "It seems particularly unconscionable . . . for outsiders surrounded by modern comforts to oppose the project and thereby condemn the people of Chad to a kind of pristine poverty," he testified. Congress was inclined to okay a project involving oil, American jobs, two huge companies, and a bold new idea. Who in Washington wanted to bet against hope?

As soon as the World Bank okayed total loans of $200 million to Exxon's project and $93 million to Chad and Cameroon, other banks joined in. The European Investment Bank threw in $41 million. And the

EXIM bank, a U.S.-taxpayer–funded bank, put in $200 million, noting that the project would "sustain US jobs" while fighting poverty in Chad and Cameroon. Two other export credit agencies tossed in another $700 million. Two commercial banks, ABN-AMRO and Crédit Agricole Indosuez, finished off the pot. The game was in play.

In Chad, Djonouma says he asked President Déby why he signed such a terrible contract, giving the country only 12 percent of oil royalties. Djonouma continues, "He had two things to say. First, he thinks oil will bring happiness to the country and people will say that Déby brought the oil. Secondly he knows that he gave everything to the oil companies. He can't exactly say that the contract was malnegotiated but that sometimes he regrets that he signed. He wanted the oil to come out of the ground so he didn't read the contract, he just signed."

Exxon and the World Bank opened themselves up to scrutiny by taking on the Chad program, but their very involvement put them in a conflicted spot: If a country's leader doesn't read the contracts he signs, does the agreement mean anything? Eighty years before, Exxon, aided by the U.S. government, had signed a contract with Venezuela's nearly illiterate dictator Gómez. Now, in a very different time, with different expectations, they had done something very similar with Déby. Caught between becoming and disappearing, Chad's state is hard to pin down, but it can appear deceptively useful to outsiders. Historian Nolutshungu mentioned that Chad's gray area allowed foreign countries, including Libya, France, and the United States, to intervene in the interests of protecting the nation's sovereignty, which he called "a hopeless paradox."

The issue of sovereignty brought up another contradiction that the Bank had to sidestep: Chad was a democracy only in name. The Bank adopted an abstract view of the connection between development and democracy. A Bank report noted that it is difficult to monitor human rights in a country where they're largely absent, concluding that "if human rights have 'significant direct economic benefits' on a Bank-financed project, they become a matter of concern to the Bank. Otherwise they don't."

I ask Djonouma what he thinks about the connection between human rights and economic development. "Now, when Americans complain about the economy the government falls. But here it's not the same case. When the economy is bad and people start complaining, the leaders will say let's bring weapons and shoot those people so they stop talking about the economy!" At this point he laughs hysterically. He is not laughing at a joke as much as he seems to be banishing demons from the room.

His cell phone rings with a loud symphonic version of the "Dance of the Sugarplum Fairies," from the Nutcracker Suite. He takes the call, then composes himself, smoothing down his dark suit. "I'm not dancing for the oil project this time. I want to wait and see."

**One day** around noon, two white Toyota trucks come careening down N'Djamena's main shopping street, cops in back blowing whistles. Soon the street is empty. Stores are locked and barred; cars are gone. Shutters are drawn. The president is coming. We can't see him, nor can anyone else. The police have cleared the streets, and it's well known that anyone caught peeking from a window may be shot. How strange to be Idriss Déby, ruler of a poor and chaotic country, and to drive through it completely depopulated, as if a neutron bomb has hit, insisting to the world that you've been democratically elected.

Déby doesn't mess around. On July 28, 2003, a car owned by a citizen named Hassan Yacine ran out of fuel near the presidential palace. When he got out to push his car, he was shot and killed by presidential guards, who were not punished.

Déby has a long history of deriding international human rights organizations as "imperialists" who "make serfs out of third-world peoples." Throughout the buildup to the start of the oil project, he continued to imprison Chadian human rights activists. There were accounts of activists who were attacked by soldiers or members of the president's clan, a northern group called the Zaghawa. The clan had full immunity. Workers at the city's electrical plant said clan members showed up waving Kalashnikovs and demanding electricity. ("It makes us nervous but we're getting used to it," the stoic lead foreman tells me.)

Déby has been a lackluster president, but he thrives at his original job: warlord. He sent troops to support the coup against the president of the neighboring Central African Republic (CAR) in March 2003. There were reports of Chadian soldiers in the Democratic Republic of the Congo as well. Déby publicly supported the government of Sudan, which had funded the rebellion that brought him to power in 1990. But his family members appeared to be supplying arms to the rebels in Darfur, who were members of Déby's ethnic group. Senior officers in the Chadian army told reporters that the elite presidential guard, all Zaghawa, were supporting the rebels. Meanwhile, Déby's troops were fighting rebel groups within his own borders.

Déby's ability to get his fingers into so many African pots had made him both unpredictable and important to the United States, which needed his cooperation in the region for the war on terror. The Al Qaeda–linked Salafist Group for Preaching and Combat, for example, was zipping around freely across the border between Niger and Chad in the northern desert. So the United States included Chad in a $500 million Pan Sahel Initiative to train African soldiers for the war on terrorism.

If the United States concentrated diplomatic energy on Chad, perhaps there would be a chance of changing Déby's behavior. But Chad, and most of Africa, is not a priority for the United States. When one of Chad's rebel groups captured an important Salafist nicknamed "Al Para," who was wanted all over Europe, the United States was unable to get its hands on him for fear of offending Déby. So the terrorist reportedly went free, though he was later imprisoned by Algeria.

And the United States got ensnared in Déby's tricks. Marines sent to train Chadian soldiers discovered that they were training the Zaghawa Presidential Guard, commanded by the president's nephew, who have the duty of maintaining the president's hold on power (and shooting people whose cars break down near the palace) rather than chasing terrorists.

**The first** check for the oil is set to arrive in late November 2003, and around the country, the optimists and the pessimists are arguing over what will become of the money. "When Chad kills an elephant it dies in Cameroon," sighs a prominent lawyer, echoing the widely held belief someone else usually benefits from Chad's efforts.

According to the pessimists, the signs are not good. By the first week in November, the president had shut down the opposition radio station and authorized the first execution in twelve years. Optimists cling to the idea that nothing goes as planned here. During that execution, for example, a bullet ricocheted and killed a member of the firing squad.

The oil project, if implemented correctly, could alter the balance of power in Chad and perhaps continue the transition to democracy for real. I meet Therese Mekombe, the head of a women's NGO and a prominent member of the college that will be responsible for disbursing the oil money. An imposing force for good, Madame Therese wears a yellow dress with a matching head wrap and gold crucifixes in her ears. She says the oil project is Chad's chance to be known for something other than war, to actually help the rest of the world.

With money from the World Bank, the college has been given its own offices, a generator, a photocopy machine, and a secretary who displays an impressive array of office supplies on her desk. (In contrast, another Chadian bureaucrat I speak with keeps his single ballpoint pen, a pair of scissors, and a stamp triple-locked inside a cabinet.) The college has nine part-time delegates, five from the government and four from civil society. It is able to okay the disbursement of money to government ministries, and it is supposed to make sure that the money reaches its target programs, but it doesn't have the ability to set budget priorities or enforce complaints against the government. In fact, the college is stretched too thin either to verify that projects are completed or to investigate spending or contracting procedures. It was in a tough spot from the moment it started, because it didn't have the money, training, or information to do its job, and it was dependent on the government (the very institution it was supposed to regulate) for funding.

"If we were in a country where laws were respected, this would be a good law," says Madame Therese. "But we're in a country where laws are not applied as they should be. Also, everyone knows that laws are just there to show the foreigners. So now everyone is asking themselves if the college will have the power as it has in the law." Madame Therese has a reputation for being above reproach and a force to be reckoned with. Clearly, she's impatient with the pessimists. "They've never seen anything like this before, so they will have to wait and see the first fruit for themselves." With each shake of her head the two gold Christs in her ears shiver on their crosses. "Those who are hoping to take the money for themselves will be afraid."

Even if the college is able to distribute the money properly, there is no guarantee that new oil fields discovered in Chad will be covered by the same agreement. Only the 1 billion barrels in the project are protected by the World Bank safeguards; there may be anywhere from 1 to 5 billion more barrels under development in Chad, and concessions have been granted to several companies, including the national oil companies of China and Taiwan. Oil exploration continues in neighboring Niger and the Central African Republic. Project boosters claim that the World Bank will pressure Chad to follow the plan, but detractors note that the agreements are voluntary.

Whenever people talk about the moral controls on the project I hear a little laugh in the back of my head. It's Venezuela's Alfredo Keller saying that corruption is not a moral problem, it's a distribution system and a way of amassing power. One of the striking things about the Chad project is how much it depends on "good behavior" and how little on good institutions.

The World Bank had a small project to build capacity in the government, but it ran far behind schedule while the building of the pipeline ran ahead of schedule. Other opportunities to use the project's structure to create a parallel framework in government were passed over. The agreement that allowed Exxon to forgo customs fees, for example, was an acknowledgment that Chad was not capable of collecting the money. The four years of pipeline construction could have been used to train a cadre of bookkeepers, inspectors, and managers to assess and collect the customs payments. At the very least, the Finance Ministry could have been electrified. When I visited, clerks were doing most of the country's accounts by hand in ledgers in a dark, chicken coop–like building.

**The World Bank's** man in Chad, Jerome Chevalier, is optimistic that the oil project will nudge Chad further toward existence. The Bank's generator is cranked and the air conditioner is blissful. Chevalier is coiffed and dressed in fitted striped shirt and trousers that for some reason remind me of the cool angular men in menthol cigarette ads from the 1970s.

We begin talking about the oil project in a predictable-enough vein. Chevalier says that the Bank and the oil companies still have a lot of clout to control the president. "Chad thought that now that they had the oil money they could be more independent but actually their degree of independence is very limited," he says. "There's a bit of peer pressure. Companies are saying 'Idriss, you're making problems. You're creating problems for us.' The signals are very clear." If we had stopped talking there, I would have been unconvinced by his arguments but convinced that that was what he thought.

But we do not stop there. Chevalier starts to talk about the electricity plant. Mismanagement, corruption, impunity, the horrible state of the current generators . . . He gets emotional. "A collection of stupidities," he calls it. "I've never seen a country with so little capacity," he proclaims, and then catches himself. "The next question is if they've shown so much incompetence, can you be sure they will do better with the oil companies?" I was going to ask that question, but now he answers it. "Of course not. There is a system in place. And this is the only way to build a constituency for building things right." But the more he talks, the more he seems to be trying to convince himself. "What do we do, tell these guys [the oil companies] to go to hell? The external NGOs don't want to fight where the problem is—in countries with big SUVs. The oil is there for fifty years, and Déby is not there forever."

Chevalier closes this conversation with a surprising acknowledgment. "I'd say it's Pascal's Wager. When Europe was emerging from the Middle Ages in the seventeenth century, Pascal said we don't know whether God exists but it's best to act as though he does." To extrapolate his bet to the Chad project, if it works it'll be great and if it doesn't, well, we won't have lost much. I respected Chevalier for his candor, but the idealistic hype around the Chad Development Project was a long way from the pragmatic cynicism of Pascal's Bet.

**Exxon says** that no one has time to talk with me. That's too bad, because I'd like to meet the anthropologist the company hired to work with communities in the oil region. She wears dresses and work boots, they say, and is known as Madame Sacrifice for her willingness to provide animals for ritual slaughter. She has been quoted as saying, "You can't change human behavior."

With the photographer, a driver, and an interpreter named Akibou who turns out to have about a thousand cousins, we set out for the oil region.

**The *CIA's World Fact Book*** says that only 2.7 percent of Chad's land is arable, but we drive south into lush lands, including flooded paddies where chartreuse rice shoots make a fuzzy line against the lavender sky.

The southern part of Chad is green and more populated than the arid, mountainous north. The French colonizers admired the northern Arab nomads and largely left them alone. They called southern Chad Le Tchad-utile, and required southerners to grow cash crops like cotton and do forced labor. Tens of thousands of Chadians died during the building of the Congo-Brazzaville railroad in the 1920s and '30s. Northerners have dominated the government since the 1970s and have continued to use southern Chad as a resource, which has created tensions. Many southerners feel that the oil project is an extension of "useful Chad."

The road south is rough, rutted dirt. Off to the sides, a few people sell charcoal and fruit, but it is mostly empty. A bottle placed near the roadway means someone nearby sells homemade alcohol. Occasional overloaded trucks or vans pass. We are waved through dozens of checkpoints, largely because of our white SUV and our white skin. The houses beside the road change from board shacks to terra-cotta cylinders with thatched roofs. Granaries stand on legs with thatched tops and homemade ropes to open them. An "Arab" town of square compounds appears, and five miles later there are oblong earth homes with giant calabash gourd plants dangling from their roofs. We wait for ten minutes while cows cross a narrow bridge. One attribute of severe underdevelopment is that things look quite beautiful and peaceful, particularly when you're riding in a car with air-conditioning and two gas tanks.

American environmentalists have fought so successfully to keep oil development out of the Arctic National Wildlife Refuge that most of us have a definite idea of what the refuge symbolizes and of the caribou and polar bears that live there unmolested. I don't think that Americans have much of a concept of Chad—most don't have any idea of where it is— and no sense of the beauty of calabashes hanging from roofs. There's friction between the global nature of the oil industry—what we don't get from the Arctic we will eventually get from Chad—and the local nature of our politics. If there is any virtue in protecting ANWR, it quickly fades in faraway Chad.

By nightfall we reach a rough hotel called The Derrick, operated by a proud woman in a white lace dress who happens to be one of translator Akibou's cousins. She takes us to her restaurant, where anti-AIDS placemats show a truck steering wheel and a condom. Cousin says that business is good. Although she enjoyed catering for the oil operation, she wasn't surprised when that job was taken over by someone related to the president. The restaurant is busy and the tables are filled with laughing men under yellow light bulbs powered by a generator.

A mark of how the project has transformed the local economy is the lack of change, as in small bills: Most business in restaurants, bars, and lodging is conducted in a system of large notes and IOUs. Small bills can be purchased at an amount above their face value. This policy is good for bars, in particular, because once a worker puts a large note on deposit at a bar, he can't take the money elsewhere. You would think that building a bank to help people save the oil windfall would have been a priority for either the government or an NGO, but I was not able to find any banks within the oil region.

Without formal credit, informal systems get the job done. A former law student now working as a caterer tells me that men get "sex loans" from prostitutes, who record their visits in notebooks and present bills on payday. He also says that the condoms Exxon gives to workers for free are taken by entrepreneurs and resold at above-market rates, when they're available. Considering that more than eight thousand workers were employed on the project during its peak, involving massive immigration of people from other parts of Chad and West Africa, the fact that sex can be purchased on credit speaks to some basic accountability and order at the local level. What this means for the spread of AIDS in Chad remains to be seen.

Men came from all over Chad to find work here. Security guards left law school or teaching jobs to make $100 a month in the oil fields, which was more than the salary of the local governor. The lure of money drained the university in N'Djamena and even closed fourteen schools in the oil region, simply because teachers quit to find work with oil. A former UN official tells me that between 2001 and 2003, the price of meat quadrupled in Bebidja, the town where most of the guards are housed. The number of motorcycles in town went from 3 to 100; 3 hotels became 120; and 8 bars became 140. A private school, the École Anglaise Génération Petrolier, made of woven-mat buildings, sprang up in an empty field, and 185 kids wearing green T-shirts with derricks on them began studying English.

The president's northern clan seems to be availing itself of useful Chad. In addition to the favoritism with contracts, I'm told that Zaghawa are grabbing land from locals. One parcel of dusty rubble doesn't look like much, but it turns out to be across from the only place in town where there's any cell phone reception. A succession of motorbikes screech up to make calls, and apparently someone in the president's clan sniffed a business opportunity.

In another town, one of the largest Chadian companies employed by the oil consortium has a head office that is empty except for a desk, chair, and computer still wrapped in packing plastic. The man they introduce as the CEO is very young, with the coloring of a northerner. He speaks no French or English or standard Arabic, which leads Akibou to conclude that he's illiterate. Completing the overall impression of a Potemkin company, the CEO wears a mustard yellow zoot suit and lensless glasses. This interview involved two translators, so I have no idea whether my questions reached him and whether the answers I received were accurate. It hardly mattered, because a cadre of better-educated southern advisors interrupted nearly all of his translated responses to say that they were wrong. The company was something out of a low-budget comedy skit, but nonetheless it had dozens of trucks driving around the oil fields. Wherever there is money in Chad, a calculus of patronage springs into action.

**The next morning** we drive toward the oil fields, which are accessible by dirt roads running through small villages surrounded by green plots of farmed land and mango, lemon, and shea trees. There are about 250 wells in the project, but few show more than a few feet of pipe branching above ground. Exxon quotes astounding numbers of material it brought to Chad: "Construction freight equal to nearly 50 Eiffel Towers or six Washington Monuments . . . 88,000 lengths of pipe from steel mills in France and Germany." But apparently the company buried most of it. In a few places, huge electric pylons march across the landscape between fenced oil facilities.

The heavy trucks that prowl the roads of the oil project travel in great roiling balls of dust, looking like giant speeding snails. Exxon, which has been criticized for not paving the roads, has responded by coating them with molasses, which it says will keep the dry dirt down. Foliage by the roads—crops, mango trees, even cows—appears to have been dusted with cocoa.

At the end of the road lies Kome Base, the center of Exxon's operations and the largest of the gathering and processing stations at the head of the pipeline. The base is surrounded by a high chain-link fence topped with barbed wire. The perimeter is lined with small sentry houses. On high posts, vapor lights flood the area, day and night. To enter the base on foot, you must walk through an intentional maze of wire fencing.

Inside the fence is a different world: mandatory breathalyzer tests, forty-one channels of cable TV, meals served in at least three levels of dining hall, and air-conditioning. A Chadian who works inside says he gets headaches when he leaves the base to visit his family. Sometimes he feels he suffers temporary amnesia and dislocation when he returns to work. "The base is America," he says.

Across the road from America sits the boom village of Kome Atan, named for the French verb *attendre*, "to wait." People wait outside the base for jobs or for food smuggled out by friends who already have jobs. The Pentecostal preachers in their crisp white shirts wait for sinners to repent. They got thirty-six at their service yesterday. Today they have only one, and his eyes are a shocking red. Behind them, men who call themselves nurses wait for people with venereal disease to buy their random collections of pills.

Kome Atan is also known as Kome Satan. Some journalists, particularly those unimpressed by the project, have said that the name comes from the prostitutes and criminals who live there, and it represents a sort of biblical fall from grace in which Edenic Chad has been invaded by Exxon's Satan. While there are a lot of prostitutes in the village—they spend their days waiting for the few hours when some of the workers leave the compound to drink in the bars—I am told there are no criminals. "This place is filled with security!" someone says, motioning me to turn around. Two thuggish men sit behind me wearing loud shirts. One has his finger buried deep in his nose.

A white man in an American-flag shirt picks his way through the neat blocks of Kome Atan. He stops to talk, saying he's in charge of all construction materials at the base and is touring the village to make sure that the scrap wood and metal has been evenly divvied up and that one or two people haven't become mobsters of scrap. With a Texan accent and a defensive stance, he squints at me skeptically and says to come find him in the bar later.

While people wait, they build houses from scrap and straw. They turn old pallets into bedsteads equipped with small locked safe boxes in the headboard. They melt down scrap beams and pour them into sand molds to make cooking pots. Yellow tape reading CAUTION can become bicycle streamers or fencing material. On the ground: millet stalks and goat shit. Piles of empty water bottles. The wrapper from Chinese playing cards, an ad for a bicycle from India, crushed cigarette packs labeled "Fine" and "Sprint."

A man in lavender robes runs toward us. He has a small Van Dyke beard and a big dimpled smile. In his hands are two cell phones and a walkie-talkie. He turns out to be Akibou's uncle, and he's eager to make us at home in the waiting village, where he keeps a small courtyard and pied-à-terre made from scrap plywood. Oncle is chief of parking for all of the project's two thousand vehicles and drivers. He is so important that five young men have attached themselves to him as his servants. They fuss around us with plastic teakettles of water to wash our hands. Oncle is fluent in nine languages. He invites us to stay for the Ramadan feast with an exuberant sweep of his arm.

Night falls. The prayer begins and in Oncle's courtyard the men bow toward Mecca.

"They gave me a room inside the base," Oncle explains during dinner, which we eat with our hands while sitting on mats, "with AC, TV, and everything. It's luxurious inside there. But I wasn't interested. I prefer to be out here with friends than in that formal place, alone, looking through the fence. Exxon would never accept this style of life. But the style of Exxon is not for me. It's no good."

"The Texans don't care about anything but work," a young man named Ali says. "Their brains are formatted this way. One month on. One month off." Another of Akibou's cousins, Ali was educated in Belgium, and he's here to run his family's water drilling business in the oil fields. He starts to talk about the rules for American workers: Some are not allowed in the bars at all, and their movements in the oil fields are monitored by GPS. "The Americans accept being treated like children, but the Chadian people don't because they don't earn that much money," he says.

As in Venezuela nearly a century before, a culture is taking root in the oil fields that reflects the boundaries and divisions most comfortable to the oil companies. Inside the camp there is one dining hall for expats, one for Chadians, and another for Filipinos, who apparently are never able to leave the base at all. The Chadians feel sorry for them.

"People hide their degrees out of fear the bosses would be afraid of your level," says a Chadian Exxon radio operator who has a master's degree in English and French literature. "When they saw the guys had master's degrees, they felt they couldn't talk like before." Of the sixteen hundred Chadians who are employed as guards in the oil fields, many are former students or teachers, though they're treated like foot soldiers by the former French paratroopers who run the guards. "We're the best-educated security force in the world," crows a former law student as he complains about working conditions.

For Oncle and Ali and most of the people in Kome Atan, the oil project is, on the whole, a good thing. They have the skills and perspective to read the situation and prosper. And they have done very well. But they have no illusions that the project will deliver anything more.

"The majors came here to earn money, not to develop!" says Ali, as he takes me to his truck to show the embroidered shirts he gives clients. They say the name of his company and ESSO, PARTNERS IN CHAD DEVELOPMENT PROJECT. "I'm part of the hypocrisy," he chokes. "I suck their dick for more contracts." He stares at me, daring me to judge him, then slams the door of the truck. "You can say what you want, but everyone here found what they were looking for. We came here for money."

**In Kome Atan,** the bars for the French are rustic and dark while the bars for the Americans are loud and bright. I find the Texan at the American bar, sitting alone, with his beer on an empty stool in front of him. He is still wearing his flag shirt. "I did two tours in Vietnam, worked in three African countries and twelve developing countries," he says. "These people are getting a bad deal. Exxon paid them $100 an acre, and some of those wells are producing 6,000 barrels a day. If they were in the U.S. they'd be millionaires even if they only got a dime a barrel."

Maybe it's not ironic that a Texan in an American-flag shirt is the first I hear who actually questions the very foundation of the petrostate in Chad—the notion that the oil belongs to the state, not the farmers who live on the land above. Despite its Americo-Venezuelan origins, the idea of the petrostate doesn't jibe with the American sense of justice and private property. These are not relationships that we would support. Simply calling this thing the Chad Development Project rather than an oil project adds a benevolent gloss to a project that would be deeply opposed at home not by environmentalists but by Texans.

The Texan is one of the recipients of the "American jobs" that justified the loan paid by U.S. taxpayers, but he is ashamed that the program is creating so few real jobs in the country. The project is not buying any food from Chad, except reportedly soft drinks. Beef is imported from South Africa, and the vegetables are frozen. Here is another opportunity to change Chad, but it was also missed. "Even in Angola," says the Texan, "the company buys food from the local farmers. But not here." He waves toward the lush fields that surround Kome Atan. "You could grow anything here." He shakes his head, saying this place is the worst he's ever been. "These people are naive."

We drive out of the oil project that night, past the young guards struggling to stay awake at their posts. The metal snaps on their black uniforms say USA. We pass the last base—an emerald city glowing through the fringe of dark trees, growing brighter until it's in front of us, its cooling towers gleaming dull silver. And finally there is only darkness until the checkpoint where the security chief gives us a big smile and a wave.

Now we are really in Chad. For the next ten miles the driver looks ahead warily, hands at eleven and one on the wheel, scanning the dark road for bandits. A U.S. State Department report on human rights in Chad for 2003 says: "Armed bandits operated on many roads, assaulting, robbing and killing travelers; some bandits were identified as active duty soldiers or deserters." Another night we barely avoid an ambush when an escaping truck waves us away. By 2004, robberies and attacks in the oil region had tripled from the year before. After dark, the state of Chad all but ceases to exist.

I asked one of the official auditors of the pipeline project whether they'd tried to measure insecurity in the oil region. "Chadians are very conscious of insecurity because they've lived with it for so long" was the unsatisfying response. The World Bank seemed to have a basic prejudice against acknowledging the things it couldn't count. And a low-level war in the oil zone was not something it wanted to enter into its spreadsheets.

In 2004 a World Bank panel recommended that the Bank stop funding oil projects by 2008 and that it should not invest in projects in countries without a rule of law. The Bank's management chose to ignore the suggestions.

But there are clear lessons here for U.S. taxpayers, who funded a program without a credible business plan (either a for-profit proposition or a development project) at the service of a policy that pursues oil projects in developing countries. And now, having effectively outsourced diplomacy to the oil company, the taxpayers will be left holding the bag if the political and socioeconomic pressures of the project become too much for Chadians to bear and a civil war starts. U.S. taxpayers will have to cover the defaulted Bank loans, pay higher prices for oil, and possibly bankroll peacekeeping troops and refugee camps. The Chad project is "limited economic rationality" at its worst. If the taxpayers who support the Bank were individuals, perhaps they'd insist that someone—Exxon, the World Bank, the Chadian government—get going on the development projects, if only to safeguard their investment. But there is no one to ask for that. It would be reasonable for taxpayers to insist that all future bank loans for oil projects include equal private investment in nonoil industries to create jobs. It would be even better to tax oil produced by the project to pay for U.S.-sponsored roads and electrification projects.

**Exxon's influence** in Chad is enormous but it has little presence beyond its fences. Early on, Exxon held community meetings where the details of compensation for land, trees, and other items were hashed out. Every agreement was photographed. There were many meetings with Exxon's anthropologist. In Chad alone, Exxon says that by mid-2003, it held 1,800 meetings with 77,000 attendees. But by the time the fields began producing oil, Exxon (or Esso), was less involved.

In Miandoum, one of the larger villages in the oil region, the chief is young and well educated. He wears a track suit and invites us to sit on the porch of his blue house while a monkey cavorts nearby. "Esso has a kind of formality. They see the world as Esso and Not Esso," he explains thoughtfully, trying to give me a picture of how relations between the corporation and the village went wrong.

Things went sour with the anthropologist, he says, because people felt she didn't deliver on her promises. The mere mention of her name now enrages villagers. "Saying her name is like throwing an angry dog in the middle of a crowd of people. Everyone runs away. She brought total confusion," he says, jumping back wide-eyed as he acts out the scene.

People in Miandoum have not done badly by the oil project. Men got short-term jobs working on the pipeline, and now there are many metal roofs. Cows, land, and mills to grind peanuts have been purchased. In the center of town, a seamstress bought sewing machines and started a business with eight employees. "When I left this village twelve years ago, the women just tied a wrap around their hips. Now they like nice dresses, fashions, shoes. . . ." She wears an elaborate pink dress and dreams that the oil zone will be known as a fashion center.

Even the chief has done well, as his secretary confides when we're alone. "In Africa, when someone catches an animal, they bring the head to the chief. And everyone here gave 10 to 15 percent of their compensation to the chief."

Miandoum's villagers may blame Exxon, but the chief is looking at the government. "The government should keep promises and should benefit the whole population with peace. But that's like a dream." In meetings, he says, the oil company and the government take turns blaming each other for things that haven't been delivered. "It's like a swing," he says, swaying his hand from one side of the porch to the other.

In a nearby town, the man in charge of a Bank-funded NGO called FACIL echoes the chief's worries. Already far behind schedule in delivering schools and health centers to villages in the oil zone, the bureaucrat in charge sits in his unelectrified office and fiddles anxiously with the papers on his desk. "If we start building the schools and then for some reason the money ends, it won't be good," he says. "We're losing time and people are losing hope. The World Bank has to do something to keep the promise to the people that the money will be used in the right way. If not, the population will blame the government and the World Bank."

But does anyone outside of Chad even care? The chief of Miandoum says he saw a lot of reporters before the oil started flowing through the pipeline. But in the last month, none. He's afraid no one wants to see the results of the grand experiment.

Perhaps because of these worries and anxieties in the oil zone, the wellheads are filled with *mystiques,* or ghosts, who wander about. The chief talks about them matter-of-factly. "These mystical things happen around the oil zone. Like you can see an old man with a heavy horse coming toward you. You see someone coming in a car and shake hands and they disappear." There are also a woman in white with a calabash and a heavy truck. What's brought the mystiques? "I asked the elders," he says, "and they say it's because of the conflict between the people and Esso—it moves all around the oil fields."

**On November** 19, 2003, shortly after the one-thirty prayer, Déby's party announces a plan to change the constitution so that Déby can have another term as president. People say this means Déby wants to be president for life and that what remains of the facade of democracy here will soon disappear.

**Oncle asks** us to come for a final Ramadan feast. He lies down with his head on one arm and tells us the story of how he quit his former job of chef du security to become chef du parking.

"When I was chef du security I was the tough guy. The villagers were always trying to steal bags of cement, wood, even generators. I had my military men on patrol, and the villagers were afraid of being shot, so they stayed away. There were stories that the people in the village of Ngalaba turned into lions and devils, but I never believed them. I always said they were telling lies.

"But one day we saw some men running at us so we turned on our headlights and the men ran back into the bush. Fifteen minutes later we saw four lions coming toward us, running to attack. I told my three men—real trained gendarmes—to fire. The mystery was that the guns were blocked and the triggers didn't move. Now the four lions were almost on us, and we figured if we got caught we'd die and never see the oil money." Oncle stops here for a round of laughter.

"There was nothing to do but move. We ran. They stole what they wanted—doors, cement, zinc, plywood, tarps. As we were running I called SOS on the radio and told my boss that we were running from lions and our guns were not responding. 'Tene bon!' said my boss. 'Hold on tight!' " Here Oncle sits up and falls down again with laughter shouting "Tene bon!"

"Everybody ran into the bush to save their own life. That night the team got back together and talked all night. We all quit. The commander became crazy because he was so fearful. He went to the homes of the lion people and said 'I'm already dead.' The security boy became crazy."

I'd heard about these lion people before, but had never heard it firsthand. How did the villagers learn to be lion people? "They got the secret from their grandparents. Those people have a bad heart. They're never satisfied." The way to scare away the lion people, says Oncle, is to hold up two pins and tell them that you're only with the oil project to get bread for your family.

I didn't really know what to make of this story, except that everyone took it seriously. The guards refused to work at the station near Ngalaba. And even the French paratroopers who ran the security guards said they believed the story. Sometimes people seemed to be describing lions, sometimes a hybrid of lion-person, and sometimes villagers. Lion and person seem to exist on a continuum that I am unfamiliar with. No particular part of a reporter's training addresses lion people.

But I did know the village of Ngalaba, which is one of the poorest in the region, and it had spooked me earlier in the week. In a small collection of mud and straw houses gathered near the woven wire of the pumping station, an extended family sat in a rough courtyard. A grandma fed a baby. An old man, femurs as visible as on an anatomy skeleton, sat mending a much-patched bicycle tire. He wore two pairs of shorts, one covering the holes in the other to make the equivalent of one whole pair. Puppies and baby goats wandered around.

The man did not look up from his bicycle tire as he said that before the oil project came, Esso's anthropologist visited often, promising that the project would improve their lives, they'd have jobs and a school. His voice is querulous. "But they only employed three people and then they fired them. They gave money to people to take the farms, but the money was too small compared to what you could make with the farms." The compensation for a mango tree, for example, was nearly $1,000, which sounds like a fantastic amount of money. But selling mangoes from the tree could bring in more than that in only two growing seasons.

Behind the old man's head, the flare at the pumping station burned, little waves of heat distorting the air around it. From where I stood it looked like a huge birthday candle, but the color and movement of its flame in the daylight made me nauseous. The project, the old man said, brought them nothing but the flare, which makes him cough. He doesn't even bother waving toward the flare. Like the Venezuelan village of Hornitos, Ngalaba had been invaded by flares, illness, and that external locus of control, helplessness.

Esso doesn't come here anymore, the old man said, and the anthropologist never even brought them so much as a package of sugar anyway. By the way, what had I brought? Nothing but the chance to tell other people your story, I replied. The old man looked down and didn't say another word.

Ngalaba worried me. Back in the States, I researched it. A 1997 report from Amnesty International described the rape of three girls in the village by government soldiers. Chad's army used rape as a way of intimidating villages, the report noted. Long before the oil started to flow, the village had paid a heavy price.

Ngalaba was also the site of an ongoing medical survey by a researcher with Johns Hopkins. The study, which started in 2001, addressed the difficulty of figuring out what was going on there. On one hand, 97 percent of the households said they didn't have enough food. On the other, people appeared to have few serious illnesses like hypertension or diabetes. Instead, they would complain about intestinal parasites week after week and toothaches for months. Hardly anyone ever went to the clinic in a nearby town, which charged 50 cents. The study, published by the World Health Organization, concluded that without access to medical care, minor illnesses became chronic, and major illnesses were not part of people's vocabulary, so they appeared not to exist. Compared to other third-world countries, the people of Ngalaba were much farther off the map of development.

The health study presents some striking images. A chart shows a baby with unremitting diarrhea for sixteen weeks. There is the thirty-one-year-old man with such bad dizziness that bending over, sitting down, and standing up was difficult. He went to the clinic once, then concluded that his condition was hopeless. And then there was the old couple who appeared to be starving to death.

I had thought of the lion people story as strange and irrational. Now here was another strange thing, irrational on one level, formal on another: a village of people dying of diarrhea that could be cured for less than a dollar sitting next to $3.7 billion of pipe.

I called Lori Leonard, the author of the study, who observed that, paradoxically, if the oil project does develop the area around the fields, it will mean that the people of Ngalaba will be able to describe themselves as sicker because they will be able to name other illnesses. She worries that the real health impacts of the oil project may never be known.

I asked her if anything good had come out of the project so far. One family used their compensation money to buy a peanut mill, which allowed them to make more money by grinding peanuts, she says. The new income made the family fight so much they split up, and the mill owner had had to move to another village. That was the success story.

I asked Leonard how the older couple were doing. She said the old man had just died. "With the compensation, the village moved to a cash economy. The old couple used to get help with their crops from work groups—mutual aid—but now they say nobody is willing to work unless they're paid in cash." Ngalaba's economy and social organization had been completely transformed in just two years. Human nature, it turns out, changes fairly easily under an onslaught.

I was glad the village of Ngalaba had the lion people, whatever they were, for defense, because they had no other recourse. Caught between one of the world's largest corporations and one of the world's weakest states, there was no institution that the villagers could depend on to protect them.

**By the end** of 2005, the oil project had given Chad $306 million in revenues, far more than anyone anticipated. In the meantime, though, Chad had moved from being the tenth poorest country in the world to the fourth. It also earned the distinction of being rated the most corrupt country in the world in 2005, tied with Bangladesh.

The college tried to do its job, but Déby replaced uncooperative members with people closer to him. The U.S. Treasury Department pulled its technical advisor out of the country in January 2004, without offering a replacement. Chad's 2004 budget allocated 40 percent of the oil revenue toward the building of a single seventy-mile road, but only 5 percent toward education and 3 toward health care. Madame Therese, the college's optimist, stood up at a dull conference on revenue transparency in London to denounce the college as "powerless," accusing both the government and Exxon of being secretive and unhelpful.

In Washington, Don Norland, the project's great champion in Congress, watched in horror, blaming the U.S. government for not reining in Déby. "This is a scandal," he told me in early 2005. "It's unconscionable to let this tyranny continue. It's so hard to convince people that this is another Guinea or Somalia." He bangs out op-eds but has trouble getting them printed now. Nobody wants to hear about another hopeless oil state.

Déby denounced the World Bank for treating Chadians like "guinea pigs" that they used to experiment with different kinds of management. In a way, he was right. He also said the revenue plan was "colonial," and violated Chad's sovereignty. Chad had been useful for outsiders, he was saying; now they would be useful for him. It was a classic lion person move—the villager who had appeared powerless suddenly had the upper hand.

A year later, in 2006, Déby remains the president of Chad, but the inhabitants of the shadow state have taken over the night. In April a rebel group reportedly backed by the government of Sudan tried to take over the capital. They were stopped by thirteen hundred French troops.

Déby and the World Bank reached an impasse because the president wanted to spend more money on arms. He changed the oil revenue law so that he could control more of the money and spend it on the military rather than development.

The task of picking up the phone and asking Déby to be nice fell to Paul Wolfowitz, who'd moved from the U.S. Department of Defense, where he'd planned the invasion of Iraq, to become president of the World Bank. He said Chad was committing a "material breach" of the agreement. In January 2006 he suspended World Bank loans.

Exxon then found itself in its own sovereignty paradox. Chad demanded that the company pay $50 million in oil royalties directly, bypassing the World Bank account. Suddenly Exxon had to decide where its contracts really lay: with Chad, the country, or with the World Bank, the guarantor. Déby upped the ante by threatening to shut down the whole project and hinting that he'd prefer that a Chinese oil company take over Exxon's concession. Worried at the prospect of losing both Déby and the entire oil project, in April Wolfowitz gave the president everything he asked for and resumed loans and oil payments.

But in the end, what else could he do? The Chad project was based on an irrationally optimistic scenario, on good intentions rather than good institutions, on roads made of molasses rather than asphalt. Just as the Bank had failed to add the invisible risks posed by bandits, amoebas, presidential relatives, Sudan, China, and lion people to its spreadsheet calculations before starting the project, it also failed to imagine how the project could go wrong or to provide tools to manage a crisis.

Wolfowitz, fresh from the failure of the U.S. war in Iraq—another venture fueled by oil, poor planning, and self-interested idealism—cynically decided to support Déby to preserve the United States' interests in Chad. But his choice reflected the United States' lackluster menu of diplomatic options when it comes to oil states. When initiatives like transparency fail, the United States' next strategy is to acquiesce to a troubled ruler while selectively engaging on topics like human rights and offering military assistance. If that doesn't work, the next strategies—sanctions and military intervention—are so much more extreme there's little likelihood they'd be used, particularly when other countries are competing for oil resources. Déby has, for the time being, found himself in the sweet spot of U.S. oil diplomacy.

The people of Chad were not the only losers in the "Pascal's Bet" of the oil project. The others were the American taxpayers who bankrolled the project and its vision of a model of a stable oil state that would keep cheap oil flowing from new territories outside the Middle East. Without that model, the United States is ever more dependent upon the Middle East, and even more in need of new diplomatic strategies to continue to compete for resources. And so, in far-off Chad, the United States had lost more than it realized it had on the table.

Since leaving N'Djamena in the fall of 2003, I often think back to an evening I spent with Besba, linguist and former parliamentarian, in a scraggly courtyard at the University of N'Djamena.

**Besba is** a tall man in his late forties with slow, deliberate movements and a beatific air of grace. As the sunset began, Besba recalled 1996, when he and 124 other people found themselves members of Chad's first-ever elected parliament. The parliamentary elections were part of a U.S.- and French-supported "transition to democracy" that legitimized the presidency Déby had obtained by coup. Nonetheless, Besba and other opposition members were excited because this was the closest Chad had ever come to democracy. He and his new friends bought books on parliamentary procedure and stayed up late arguing about what democracy meant.

In the afterglow of the setting sun, Besba described in a very evenhanded and unemotional way how democracy foundered. Besba fell silent, and the weird intimacy of sitting near a stranger in total darkness set in. The courtyard smelled slightly of vanilla. When he began talking again, his calm evenness was gone: He was angry at the lack of electricity at the university, frustrated by the pointlessness of assigning homework to students who had no light to study by. This was all symptomatic of far bigger plans rendered useless by mismanagement, corruption, and impunity.

Besba remembered when the oil project was first proposed in parliament. When he and opposition members asked, they discovered the government had long ago signed the deal with Exxon. "The debate was around whether Chadians were going to get enough of a role in the oil production," he said. "We thought it was not enough, but it was take it or leave it."

The parliamentarians had spent most of their lives at war. They'd seen oil start wars in Sudan and Libya. They were fearful. They did not see the oil project as a bet about a 49 percent chance of failure or a bet against hope; they saw this as a bet of their whole world.

Before the vote, someone stood up and made a speech, which Besba rendered this way: "In my area there is a certain type of bird. When you see that bird in the forest, you know you will lose either your mum or your dad. This is the case with petroleum. We need it. We need the money. But it should bring us more than what we have. Oil means that something will change—you cannot choose if it's your mother or your father who will die. Something bad will happen whether you like it or not."

# 9 ⁞ IRAN  *REVISITING A MINI-EPIC BATTLE*

Early on the morning of April 18, 1988, Bosun's Mate Third Class Anthony Rodriguez got up and began to go about his business on the deck of the USS *Wainwright*, which was sitting in the Persian Gulf, preparing to shoot an Iranian oil platform. He did some deck maintenance, worked a bit on refueling the ship's small boats, and got the deck ready for the launch of the helicopter. Standard operation. It was one of the aspects of navy life he loved. "On the *Wainwright*, everybody was close knit. Nobody was stabbing anybody in the back. No brown nosers. Everybody knew their job and the next."

Rodriguez had been in the navy for two years. The cold war was waning, and he'd spent time off various coasts. "Sitting. Doing whatever a cruiser does." Then his ship got sent to the Persian Gulf, where the eight-year-long Iran-Iraq war was still going on. Officially neutral, the United States "tilted" toward Iraq, and was escorting Kuwaiti oil tankers (which had been reflagged as American ships) through the Gulf to protect them from attacks. Rodriguez hated the Gulf. Ungodly hot. The surface of the water was slimy, he said, and "infested with sea snakes who'll bite you between the fingers and toes."

Four days earlier, an Iranian mine had torn a massive hole in the U.S. frigate *Samuel B. Roberts*. This morning's attack, named Operation Praying Mantis, would be retribution against the Iranians for that mine. Rodriguez's expectations were simple: "Grease up the missile launchers. It's gonna be an early Fourth of July." He had no idea he was about to be in the middle of the largest naval battle since World War II.

At 7:55 A.M., Operation Praying Mantis began. The *Wainwright* and two smaller ships took positions around an oil platform and announced in Farsi and English that the crew had five minutes to get off.

The crew asked for extra time to exit. The U.S. captain gave them about a half hour.

At 8:30 the three ships began firing a thousand five-inch bullets at the platform. A single five-inch shell, Rodriguez says, is approximately three feet long and weighs eighty pounds. "They make a nice big hole." After a few minutes of shooting, a shell hit a compressed gas tank, setting the whole platform ablaze, and whoever remained on it jumped off. (One American account says that incendiary ammunition was loaded in the guns by accident, and the platform fire was a mistake.)

Back on the *Wainwright*, by around 9:00 Rodriguez figured the day's work was over. "We were relaxing, just sitting there protecting Iraq from Iran," he says, when an Iranian warship came into view.

Rodriguez heard someone scream "video separation," which meant a missile had been fired at the *Wainwright*. The missile heading toward the ship was a U.S.-made Harpoon, which had been sold to Iran when the Shah was in power. Soon Rodriguez heard it. "Like an F-14 taking off a flight deck, a loud thunderous sound. Louder than an airplane engine. We had orders to 'embrace the shock,' which meant you just held on to anything massive. Wow. There's only so much you can grab onto in that dispensation of time. I grabbed some irons on the bulkhead. If it hit we were going to die because we were under the bridge, so I thought about the things I should have done. It was my life flashing—the way you're gonna go. You know you're going to die. Definitely."

Harpoons travel close to the surface of the water, lock on to their targets, and then rise quickly to come down on them. The *Wainwright* turned sideways to make itself into a smaller target and tossed some metallic confetti into the air to fool the missile, which dropped into the water nearby. Then the *Wainwright* and the other U.S. ships began firing, and soon the Iranian warship was on fire and sinking.

Although the Persian Gulf holds more than half of the world's oil reserves, it's just twice the size of Lake Erie. As the day wore on, fights broke out all over the Gulf. As planned, three other U.S. ships attacked another oil platform named Salman. Then small Iranian speedboats shot at a U.S. helicopter, attacked a U.S. supply boat, and peppered an oil platform off the coast of Abu Dhabi with grenades and machine gun–fire for four hours. An Iranian frigate fired on three navy jets, which shot it with laser-guided bombs before a U.S. warship sank it with a Harpoon missile. The *Wainwright*, though, wasn't done. Two Iranian F-4 Phantom jets (also purchased from the United States when the Shah was in power) came in to attack, and the *Wainwright* shot at them. Then U.S. bombers attacked one of Iran's largest warships. By the end of the day this accidental battle had become, according to naval historian Craig Symonds, one of the most influential naval engagements in U.S. history, right up there with the Battle of Midway. And it was the beginning of a complex, often contradictory U.S. military involvement in the Gulf.

Ironically, the United States attacked the oil platforms specifically to *avoid* getting into a battle with the Iranians. Admiral William J. Crowe, the chairman of the Joint Chiefs of Staff, had hoped to exact retribution for the mining of the *Roberts* by attacking an Iranian warship, but President Reagan and his advisors were, in his opinion, "dealing in perceptions; what they really wanted was to make something out of nothing. That meant striking a blow that would not hurt the Iranians so much that they would be moved to escalate." Crowe writes that Reagan was not "overly familiar with the region or terribly inquisitive about it." Even though both Iran and Iraq were attacking tankers, he was attracted to the moral picture painted by his aides of rogue teams of Iranians attacking ships like pirates. In Crowe's view, for Reagan, "The Gulf presented a vivid picture of right versus wrong."

The American public opposed sending U.S. troops to the Gulf, so Reagan used his folksy way to pitch the move as the answer to fears of another oil crisis, even though the region sent most of its oil to Europe and Japan. "I'm determined that our national economy will never again be held captive, that we will not return to the days of gas lines, shortages, economic dislocation, and international humiliation. Mark this point well: The use of the sea lanes of the Gulf will not be dictated by the Iranians," the president said.

The nine-hour fight ended with two oil platforms burned, wiping out 150,000 barrels of oil production a day, six Iranian ships sunk or damaged, one Iranian plane down, and at least fifteen Iranians dead and twenty-nine wounded. Half of Iran's navy had been destroyed. A helicopter accident killed two Americans.

Rodriguez hoped to exit the Gulf immediately. "Leaving. That was the biggest lie I told myself," he says, "I've been back four more times." In

1990 he was back fighting against Iraq, in the first Gulf War. "I was shocked to be fighting Iraq," he says, "after protecting them from Iran in '88. If we'd predicted that, we should have let the Iranians take 'em down." He was back twice later in the decade on the missions to enforce the sanctions and no-fly zones on Iraq. And then in 2003 he was back for the second Iraq invasion, this time moving hovercraft from sea to shore, ferrying weapons, cargo, and troops. Operation Praying Mantis was the beginning of it all, but it remains, in Rodriguez's words, "one of those mini–epic battles not many people know about."

In the years since 1988, the U.S. military presence in the Gulf has grown from nothing, to $50 billion a year for the 1990s, to a full-scale occupation costing more than $132 billion a year in 2005. By one estimate, the hidden costs of defense and import spending are the equivalent of an extra $5 for every gallon of imported gasoline, a cost that doesn't show up at American gas pumps.

And while Americans question the morality of using force to secure oil supply, not many question whether it is an effective strategy or worth the money. Whether leaders echo Reagan and say force is necessary to prevent long lines for gasoline, or Iraq war protesters shout "No Blood for Oil," the underlying assumption is that displaying force creates an atmosphere of overwhelming power and hegemony, preventing attacks on oil facilities and supply lines. I wondered if this was true, and if it remains a valid assumption as the market for oil has changed.

I came across a mention of Operation Praying Mantis in a 2003 ruling by the International Court of Justice. The ruling, which didn't get much press in the United States, determined that the United States' destruction of the Iranian platforms was not justifiable as self-defense. I thought it was strange that I had never heard of this bizarre one-day war with Iran. I wondered why the attacks weren't better known. "It still hasn't sunk in," said Gary Sick, head of the Middle East Institute at Columbia University and a former member of the National Security Council under Presidents Ford, Carter, and Reagan. "From the platforms, there was an unbroken arc of increasing involvement [in the Middle East] which is not understood by 99 percent of the public. At the end of the game the U.S. has become a Persian Gulf power. We don't have a doctrine, so we've improvised and dug ourselves into a very deep hole. When we talk about the Gulf now, we're not talking about some exotic foreign place, now we're talking about our neighbors."

And while every American politician promises to somehow change our relationship with the countries of the Gulf, or its oil, and pack up the troops for good, Rodriguez has reached the opposite conclusion. When his daughters said they wanted to join the military, he warned them they'd spend their lives in the Gulf.

**The flight** from Frankfurt to Tehran is full, and the man in the seat next to me is an Iranian businessman with Canadian citizenship. He is young, hip, and well connected. He wears cologne. He leans toward me conspiratorially and begins to talk. Cement in Tehran, he says, is all substandard. The buildings are crumbling, but you can't see it yet. The children of government officials throw naked pool parties with cocaine and heroin in the hills. The mullahs, he says, have lost their legitimacy and are just in it for the money. (I was traveling in September 2004, before the election of conservative president Mahmoud Ahmadinejad.) As his conspiracies get better and more complex, the angle of the businessman's body changes, tilting with intensity; his eyes stop moving and become fixed. A secret plan, he says, calls for cutting up Saudi Arabia in three parts. Before I even set foot on the ground I am deep in Iran's Conspirastan, a parallel world of explanations for the puzzle of oil, religion, and power in the Persian Gulf.

The businessman remembers the platform attacks well, as does everyone else in Iran, where the bombing was perceived as the first half of a "message" that the United States intended to make sure Iran lost the Iran-Iraq war. The second half of the message arrived a few months later, when American forces accidentally shot down an Iranian passenger jet, killing 290 passengers. Many Iranians believe that the United States urged Saddam to invade Iran in the first place in 1980 and that the shooting down of the passenger jet was no accident. In Iran, the external locus of control of the oil state has been turned into an art form, even a way of life. And often enough, the conspiracy theories have some truth to them.

The cabin lights dim and the plane is almost dark. The businessman becomes less suave and more reflective. During the war, he says, Iraq bombed Tehran nightly. At school the teachers would take roll call, and if someone didn't show up for a few days, they'd assume he and his family were dead. These stories of war and death are the twin of the more theatrical conspiracies, and they are told in an undertone.

**The United States** has not had formal diplomatic relations with Iran since the Islamic Revolution of 1979 and the hostage crisis that followed. In the meantime, Iran has resumed its place as a regional power and become a model for Muslims aspiring to form Islamic states. A brief thaw when Iran helped the United States pursue Al Qaeda and the Taliban in Afghanistan ended abruptly when President Bush announced that Iran was part of the "axis of evil." After the United States invaded Iraq in 2003, some hinted that Iran was next. A British official told *Newsweek,* "Everyone wants to go to Baghdad. Real men want to go to Tehran."

I also wanted to go to Iran, and because the United States lost its case in the International Court of Justice, I thought I had a chance of getting a visa. I wrote a letter to the state oil company asking if I could visit.

I didn't hear back, so I wrote to a journalist friend of a friend. He forwarded my note to someone, who forwarded it to someone else. Eventually I received an e-mail saying "Hi. I will help you." My correspondent was an Iranian oil reporter, she said, and I should call her on her cell phone after midnight my time. I did. She had a young voice. I imagined her in full black head-to-toe robes. Did she know what she was getting into? Yeah, she said in slangy but very correct English, she'd help me.

And that was the start of the fax war. Every morning, for several months, I'd get up early to check my e-mail and there would be a note from Aresu Eqbali, telling me to send a fax to a certain individual, at the following number, saying x, y, and z.

One morning I found a note advising me to write a "spicy" letter to an official in the Ministry of Islamic Guidance. I didn't know what to make of this. Perhaps this Aresu person had a strange grasp of English? "Spicy" to me meant flirty. Writing a flirty letter to some stone-faced gentleman in the Ministry of Islamic Guidance didn't jibe with my image of the Islamic Republic of Iran, but obviously I had a lot to learn about the chemistry of religion, power, and sex. I read and reread her note and decided, finally, to write a funny, self-deprecating letter to the official. The next morning I got a note from Aresu reading "That's enough spice to turn Mr. X on!" The spicyish letter succeeded in the sense that the official said I wasn't any concern of his; we needed to start writing letters to the Ministry of Petroleum. On August 10, 2004, Aresu sent an e-mail: "The oil ministry has given a big green light to your plan." So I bought a plane ticket.

**As the flight** into Tehran comes to an end, the attendant says it is time for women to cover up. Around me women stand and rummage about for their coats and scarves. Their coats turn them into smooth monochrome hourglasses, like Swedish-modern pepper grinders. Their scarves are sheer, chic, and in pastels that match their eye shadow. They smile at me with exaggerated pity—I am wearing a giant blue dress from a discount outlet and a thick piece of fabric is pinned under my chin. No one will mistake me for spicy.

The airport is faded and worn. At customs there hang pictures of Ayatollah Khomeini, bearded and surrounded by roses, a mildly disapproving grandfather. Next to him, his uncharismatic understudy, Ayatollah Khamenei stares deferentially to the side in his thick plastic glasses. The axis of evil as fussy comedic duo. And as for having to wear the scarf and covering, someone has tried to sweeten it up. An English-language sign in the airport reads: DEAR LADIES: HIJAB IS LIKE NACRE AND WOMAN IS AS PEARL.

I have miscalculated gravely in the conservative direction. A photographer I've e-mailed with for a few years meets me outside the airport. "Come from a very strict family, do you?" says her husband, when he sees my outfit. Changing money, I say my phrasebook-practiced Farsi word for "thank you," and get a giggle and "We say merci."

I check into a small, slumping hotel full of visiting Iraqis and Kurds. The hallways smell delicately of sweat and rosewater. My room is taller than it is wide, with a chandelier and a fluorescent light that will not turn off, strobing madly through the night. In the morning the hotel serves breakfast on an old Grundig radio cabinet. Pickles, farmer cheese, jam, watermelon, yogurt, and dried fruit have been gathered around a portable fountain decorated with little naked cherubs. Crackers sit where the record player used to be. "Frozen in time" doesn't do it justice; it's more accurate to say that it's caught in a sort of time slush, swishing around in an era of its own.

I call Aresu, who says, faintly, that she is on deadline on a nuclear story. I should try her later. "Are you chilled out yet?" she says. I spend a moment under the chandelier wondering if we will really go to the platforms and if I will even meet her.

I turn on the Shahab TV. A field of pansies gives way to a peaceful view of morning glories on a radio transmission tower in the Alps, accompanied by hopped-up electronic music. The music makes the pansies seethe, as though the flowers are quietly ticking toward an explosion. On the next channel, Ayatollah Khamenei in his glasses and sandals watches an endless parade of soldiers marching in a dry, tan-colored place.

According to a next day's account in the English-language *Iran News*, the supreme leader is watching Revolutionary Guards fire short and medium-range missiles on "fictitious enemy positions in western borders." The medium-range missile, the reporter says, is probably similar to something fired last month, which could carry a 2,250-pound warhead. The article continues, in a dry style bordering on the sarcastic: "Although the missile has been paraded with the banner 'Israel should be wiped off the map,' Iran says it is purely defensive."

Iran insists that it is developing nuclear fuel for electric power yet its belligerent pose suggests the country would like the world to think it is developing a weapon. British estimates say if Iran "threw caution to the wind," it could produce a single nuclear weapon by 2010. Official U.S. calculations are that Iran may have something ready by 2008 to 2010. The most strident American commentators, though, have stated in the past that a weapon would be ready as early as fall of 2003, although that deadline has obviously passed.

In Iran, the nuclear weapon is seen as the ultimate sovereignty insurance. Surrounded by U.S.-occupied Afghanistan and Iraq, nuclear-armed Pakistan, and Azerbaijan, which is considering housing U.S. military bases, Iran knows it needs to appear strong to avoid being perceived as weak. "Iran's quest for nuclear weapons does not stem from irrational ideological postulations, but from a judicious attempt to craft a viable deterrent posture against a range of threats," Iran expert Ray

Takeyh told members of Congress. The United States invaded Iraq, while it left North Korea, which possesses nuclear weapons, alone, he noted. "The contrasting fates of Iraq and North Korea certainly elevate the significance of nuclear weapons in the Iranian clerical cosmology."

The nuclear weapon also provides symbolic development, the long-promised modernization from the oil money. The Shah himself longed for a nuclear power program and tried to purchase first-world accoutrements.[1] But Iran's oil wealth developed the country's aspirations without developing its infrastructure. So the quest for nuclear weapons comes off as surreal here, where a headline from the Tehran *Times* reads: "Iran lacks technology to produce black chador." In a country with dozens of millions of chador wearers, the ability to produce black dye and fabric (which requires petrochemicals, one thing Iran has lots of) is simply not there. A nearby article complains that government mismanagement is about to bankrupt the entire powdered detergent industry.

**I am in touch** with a man who is on the outs with the regime. We arrange to meet. He is tall, wearing a plaid shirt. He is a larger-than-life charismatic character, uncomfortable with his invisibility and yet somehow committed to it. "There is another world from here which is only five minutes away," he says of the difference between being ignored and being punished. We go to a dark restaurant, heavily hung with brocade that seems to date from the Shah's time. A buffet holds fifteen kinds of salad and about 75 percent of them are pink, giving the somber hall a festive cast. He encourages me to take out my notebook, and even though the sauce-stained waiters are studiously inattentive, I still feel paranoid on his behalf.

"You can't sort this out anymore—politics, Middle East, oil, super-powers, demographics in the U.S. and the Middle East, religion. . . . All

---

[1] In the 1970s the United States supported the Shah's plan to build nuclear power plants and a giant grid. The program would have brought $6 billion (in 1976 dollars) to U.S. companies. The Shah maintained that oil was a resource too precious to be wasted on producing domestic electricity and should be reserved for export. According to a report in the *Washington Post*, President Ford not only bought the argument, but okayed an agreement to sell Tehran a nuclear reprocessing facility for plutonium in 1976, with the support of Henry Kissinger. Ford's chief of staff at the time was Donald Rumsfeld, who was succeeded by Dick Cheney.

of this is in one bag. An enormous bag. And it's very difficult to understand! Even if you put your head in the bag." He mimes sticking his head nervously into a bag—possibly one filled with fighting cats—and pops back out, laughing. "Even if you're an expert in it!"

He holds the United States responsible for the way Iran is today. When America engineered a coup against the nationalist prime minister, Mohammed Mossadeq, and put the Shah back in power in 1953, he says, we took responsibility for Iran's fate. "In 1953 you Americans were very novice in the game—you couldn't care less about Iran. But how could you lose it? You had it in your hands!"

Some backstory: The U.S. interest in the Gulf is partly the fallout of the Black Giant—the enormous East Texas oil field that depleted rapidly in the 1930s. As World War II ended, Washington planners estimated that the country had enough oil to last just two years in a war with the Soviet Union. When FDR made an agreement with King Saud in 1945 allowing American companies to drill for oil in Saudi Arabia, his advisors saw it as a sort of New Deal for Saudi Arabia and the Gulf, leading to "increased purchasing power and greater economic and political stability." And oil, of course.

The first crisis of the cold war appeared in Iran in 1946, when the Soviets didn't withdraw after the end of World War II and began to talk about "liberating" Tehran from capitalism. The United States hastily made agreements with the Shah of Iran. It was a kind of chess move: Block the Soviet Union, protect the oil, and feed energy to rebuilding Europe and Japan. Saudi Arabia and Iran—one with lots of oil, the other with geostrategic heft—became the "twin pillars" of U.S. influence in the Gulf.

As the British Empire faded after World War II, the United States took up its position in the Middle East but without the imperialist baggage. The United States preferred to think of the Gulf in business terms, as a complicated political and economic machine—something like a fast food franchise that sold oil from a drive-through window. Pull up, get your oil, and move on out.

Both the Saudis and the Iranians initially interpreted these arrangements as proof that the United States backed up their sovereignty. However, the Americans had secret plans from 1949 onward to destroy Saudi oil fields if the Soviets invaded, a move that seemed to violate the spirit of the agreement. According to articles written by Steve Everly of the Kansas City *Star*, the United States went so far as to ship explosives to the kingdom. Some oil company executives stored them under their beds for years.

In Iran, the United States saw itself as the corporate manager of the drive-through window, entitled to choose its leaders. A Central Intelligence Agency (CIA) history of the 1953 coup describes Mossadeq the way one might evaluate a CEO gone amok: "governed by irresponsible policies based on emotion." The CIA planned to replace the government with "one which would govern Iran according to constructive policies," but they were particularly concerned about the country's oil. With American help, the Shah of Iran returned to the throne to become the policeman in the Gulf, and the United States gave him as many weapons as he could write checks for.

The Shah was an anxious man with a big wallet, and in the early 1970s he spent $11 billion buying American military hardware (including the Harpoon missiles later fired at the *Wainwright*) and $100 million on a four-day celebration of modern Iran, where his wife wore a ten-pound crown containing emeralds the size of golf balls. His secret police and hostility to religious authorities made him unpopular within the country, and the United States knew it. A 1977 U.S. Senate report cautioned that America should not be "sanguine" about the Shah's hold on power, and should consider the degree to which the United States was implicated in executions performed by his intelligence services.

In the 1970s, my host says, he talked with people from the U.S. embassy, but they didn't want to listen to the rumors of revolution. He thinks the United States would have forced the Shah to reform if Kennedy hadn't been assassinated. ("See, we lost something in Dallas too!") Or if Nixon hadn't gotten embroiled in Watergate. ("Watergate was not just some stick behind some door!")

"The U.S. has made a lot of mistakes, and we have really paid," he says. "There were so many snowflakes and suddenly the snowball is falling and there is nothing you can do." In his telling, there were many opportunities to avoid this particular Iran, and the United States is at fault for not taking advantage of them.

Only in Iran is the external locus of control so strong that people insist that their own revolution is America's doing. But even though Iran and the United States have had no relationship for decades, American power and interest in the region's oil loom large.

"The mentality of looking for solutions outside the country runs very deep here. It comes from a history of conspiracy. There's a feeling that major change will not happen here without external intervention," an Iranian commentator tells me. "It sometimes seems irrelevant if we're talking with other countries in the Gulf."

My host worries that as oil fields around the world are depleted, leaving the bulk of supplies in the Middle East, the world's wrath will turn here. "Things will start to get crunchy," he says with a grin. "If I'm right, finding oil will be an enormous problem for the U.S. suburbia," he says. "They are the most important socioeconomic community on this planet, and they are not going to take the destruction of their way of life lying down. They have an enormous power to change American politics—everything is possible. Maybe even an end to democracy. Forget about nuclear weapons and terrorism. I am very worried about the explosive power of panicked suburbia." Will the United States dissolve without oil? This wacky idea has its mirror opposite in the American belief that a military assault will instantaneously unravel Iran's theocratic power structure.

**Gasoline holds** Iran together. Tehran has 3 million cars, all burning copious quantities of fuel that costs just 34 cents a gallon. Gasoline is at the heart of the government's pact with the governed, and prices must be kept artificially low. Because it is so cheap, gasoline consumption is rising by 10 percent a year, contributing to the city's spectacular traffic jams. In a country with unemployment of as much as 15 to 30 percent, cars and gasoline are a balm for life with few prospects. And while many aspects of life in the city are repressive, traffic rules are not.

The city of Tehran may be the world's greatest stock-car circuit. Its French-inspired boulevards suggest a city for promenades, but walking in Tehran is dying a dozen deaths a day. Cars roar along without slowing for pedestrians. To cross the street, you must run from lane to lane, pausing on the dotted lines, turning feet out in a ballet stance to avoid losing your toes. When the cars clear, you conquer the next lane and so on until you can stand on the median and contemplate the next assault. The air is bluish with exhaust. Digital readouts on traffic lights count down from ninety seconds to zero, making every stoplight the equivalent of a countdown to takeoff. By the time the light turns green, many cars have already crept, catlike, into the intersection, and now they leap as if pouncing on something on the other side.

The soothing power of gasoline can't be underestimated. A friend takes me to visit a female race car driver named Laleh, who lives in a wealthy neighborhood in her parents' house, which includes an indoor swimming pool and a lot of gilded Louis XIV furniture. She's studied in London and raced in Iran and elsewhere in the Gulf but her life in Tehran is a sort of gilded tragedy, sweetened mainly by gasoline. At twenty-eight, Laleh says she doubts happiness will ever be hers. She has a master's in mechanical engineering, an interest in painting, and facial features sculpted by good plastic surgeons.

It's night when she gets behind the wheel of her rally car, wearing elegant gloves and letting her head scarf slip back as she sets off to conquer the outer rings of Tehran. We screech around garbage trucks and slide between overloaded motorcycles as Laleh accelerates happily. With one gloved hand on the wheel and the other on her cell phone, she cuts off taxis, slips left and then right, surmounting all obstacles, surpassing them, leaving them nothing but exhaust. She is finally happy when she is ignoring my screams. "Enjoy till you die," she says, without nearly enough irony.

No government is brave enough—or has enough popular support— to do away with cheap gas, but it's gradually killing the Iranian state. Iran consumes far more gasoline than its overburdened refineries can produce, which puts the government in the awkward position of having to export crude oil and then pay to reimport gasoline from foreign refineries, at a cost of $4.7 billion in 2004. As oil prices rise, these costs increase, and Iran's ability to export oil is further crimped. Conservative members of parliament want to reduce the price even further, to about 11 cents per gallon, perhaps as compensation for the dreadful economy and further social restrictions, but they are opposed by government economists, who point out that oil subsidies are eating up 10 percent of Iran's GDP. Every year, five thousand people die of smog-related sickness in Tehran alone.

~~~~~~~~~

**Oil is Iran's** universal solvent, Aresu says when I finally meet her in a European-style coffee shop in the north of the city. In her mid-thirties, she wears an olive drab coat and head scarf and deep red lipstick, which reflects her stealthy sense of humor. Disguised as a staid oil reporter, she's got a mocking sense of the absurd, anger, and a taste for old-fashioned, mildly dirty jokes.

Oil is the answer to every question and every dilemma in Iran, she says. Whatever problems oil brings, it solves more. The nuclear issue? Oil will solve it. The human rights issue over a Canadian journalist who was killed in prison? Oil, she says, will somehow provide a way out. Oil ultimately trumps ideology, religion, and bone-headed stubbornness; it keeps Iran engaged with the world no matter how much some in power might prefer to retreat.

Oil may be all-powerful, but Aresu's not caught in the romance of it. A former German literature grad student, she's not even particularly interested in oil. What she likes is the ironic situations it creates, as if oil were a wily character in a cartoon who simultaneously flatters and takes the air out of the pompous. Oil makes everyone look ridiculous. She begins to laugh, pointing over my head at a bookshelf on the café wall. On the shelf there's a trashy novel from the 1970s titled *Oil* by Jonathan Black. The cover reads: *Oil Is Life, Death, Money, Power, Sex. Without It, Industrial Civilization Shudders to a Standstill—and Rots.*

Oil accounts for 80 to 90 percent of Iran's export earnings and nearly half of the government's budget. Distribution of that money sews together a diverse crowd of the very religious, the very poor, the very rich, the angry, and the opportunistic. And the particular way these relationships are formed makes Iran's power structure extremely resilient. Here the petrostate triangle has expanded to include many other institutions with religious and military authority, so that it no longer resembles a triangle at all, but looks more like a well-woven rug.

Fifty-eight percent of the national budget is allocated to organizations called *bonyads*, which are tax-exempt and are headed by clerics who answer to the supreme leader. After the revolution, the bonyads were run by religious leaders as charitable foundations redistributing profits obtained from "blood-sucking capitalists." Over the years they've grown into huge business empires with social leverage—the ability to start protests, provide patronage, set policy, and repress dissent. They combine economic heft—controlling as much as 20 percent of Iran's GDP—with political power and religious authority. The bonyads have been described as a "shadow government," and the U.S. State Department has said that it suspects some of them of supporting terrorism in other countries.

"Iran is much more corrupt than you can possibly imagine," says an Iranian businessman who sets up deals between Iranian and foreign firms. (He invites me to his office and then suggests that I remove my head scarf. I don't; it somehow feels like trying to do an interview in a bikini. Iran has messed with my head.) Anyway, on with the corruption. "Unbelievably corrupt. The problem with the National Iranian Oil Company is its mafia has been there a long time, and through marriages the number of people involved has risen and the number of factions involved are fighting for power among themselves." He says that old-money families have married into clerical families, which means that the two groups now have common interests.

He describes extraordinary schemes in which the ruling families and foundations skim money off the top of foreign contracts. It's not only that they take the money, it's that contracts are given to foreign companies whose agents promise the most. As a result, billions are wasted and many projects are never completed. He goes through a litany of specific projects detailing payouts to either religious groups or intelligence officials. His story is believable largely because it mirrors other reports. When the Norwegian state oil company Statoil tried to invest in Iran's oil industry in 2001, it was told to talk with the son of influential politician Rafsanjani, who suggested that they pay a "success fee" to him or to a bonyad. Finally, Statoil agreed to fund Rafsanjani Jr. as a "consultant" through a third party in London for eleven years for $15 million.

As in Venezuela, corruption is an efficient distribution mechanism. Iranian government economists have tried to quantify the size of the unofficial economy, and it seems that virtually everyone is involved. Most households appear to spend more than they officially earn. By some estimates, the underground economy has grown since the mid-1990s to be almost a third the size of the official one.

The businessman laughs at the idea of the mullahs losing power, which some in the United States believe is imminent. He explains that the dispersal of power and money through the religious and financial networks, with the added strength of intermarriage, will stand in the way of a sudden revolution. "If you shook the tree, the birds would fly off and then resettle on the tree." What is needed, he says, is a more fundamental change within Iran to dismantle corruption.

Middle-class Iranians believe that the lower classes are loyal to the mullahs for political and economic reasons. They sense that the religious lower classes are on a mission of revenge. "The mullahs had a complex. They were poor and they never had access to money or beautiful women so they had complexes," explains the businessman. The poor are the main recipients of $6.5 billion in subsidized bread, rice, sugar, cooking oil, cheese, and prescription drugs. No one expects the relationship to change anytime soon. "Once you have tasted power—it's delicious—why would you give it up?" a government advisor tells me.

This mix of piety, anger, and oil subsidies makes the regime particularly invasive. One woman tells me of getting arrested at a party when she was in her late twenties. (There was no booze, only men.) She was jailed and eventually caned. The lasher, she says, had been certified as a properly Islamic lasher, and wielded the weapon in accordance with Islamic law while holding a Koran in her armpit. The middle-class party girls tried to shove cardboard under their clothes to avoid the pain, but the lasher was diligent. She started at the feet and worked her way up. The woman says she had welts for a few days. "The lasher was an angry woman and probably from the lower class. She was someone who got things from the revolution," the woman explains. She starts to laugh a horrible laugh, mocking herself for having been foolish enough to feel pain, for having accepted the lash's humiliating personalization of power.

The desires of the 70 percent of Iranians under thirty are unknown. They are the ones who face the worst of the sluggish economy and joblessness. Religion, allegedly, is no longer much of a balm, and that seems to be borne out on the day an acquaintance takes me to visit Khomeini's shrine in the south of Tehran. Few people are there, and those who are seem to be mostly Afghan refugees, even though it's a holiday. On the way back, we ride in the family car of the subway, which is almost completely filled with young men. They squash together and try not to fall on the women on the bench seats. The car feels oppressive, and it seems we're being examined by a hundred X-ray eyes. My friend responds to the atmosphere by translating the juicy parts from "A Mullah's Advice to Brides on Their Wedding Nights" into English. (Do not have sex standing up, or from the rear. But if your husband asks you to do it in a saddle, you must agree. More obscurely: Avoid sour apples.) She is snickering a wicked, angry laugh, full of scorn. I wonder if she's trying to get us into a fight. And I wonder what the young men around us are thinking. The moment they stop seething and begin to act will be the start of Iran's future.

The poor may want change. When many democracy advocates sat out the 2005 presidential election, the poor elected conservative Mahmoud Ahmadinejad, who offered the Iranian version of "a chicken in every pot"—which was to put the "petroleum on their tables" and fight corruption. Ahmadinejad's first attempts at taking power stuttered as parliament repeatedly rejected his nominees for oil minister. Unable to deliver the petroleum to the tables, he became inflammatory, delivering instead anti-American and anti-Israeli rhetoric, aggressive nuclear posturing, and dreamy religious visions. He emerged as the quintessential petro-politician, a cabaret singer belting out all the old oil-state favorites. But beneath the noise (and it is loud and frightening), he is the first Iranian leader in more than twenty-five years to attempt to start diplomatic relations with the United States. A page out of Chávez's book.

**Aresu picks** me up in her car wearing a daredevil grin. We're off across Tehran in search of a former oil executive to interview. She weaves and dodges in traffic, driving through intersections while talking on the cell phone, as we suck down lots and lots of exhaust. We come up empty, finally, in the search for the executive, and I wonder again whether I'll ever get close to those oil platforms. We head to a café where young women "volunteer" to wait tables because women are forbidden to waitress. They wear full dark cowls and don't appear to be having much fun, but what do I know? Aresu says that women have become more reckless as the reforms of the late 1990s went nowhere. "So you put me in jail and you let me out! So?"

We end up at the National Iranian Oil Company (NIOC) headquarters, a 1970s high-rise where posters celebrate the oil workers as martyrs of the Iran-Iraq war. We are greeted by the representative from the PR department who received my early-morning faxes. She is a small woman wearing a black cowl that reaches past her stomach and a gray twill coat that almost brushes the ground, leaving just a tip of Band-Aid–colored shoe peeking out.

She seems anxious, and I wonder if she regrets her role in letting me into the country. She quickly hustles us in to see Mostafa Zeinoddin, the NIOC lawyer who handled the court case against the United States for bombing the oil platforms. He's a big man with a serious expression that occasionally breaks into a pumpkin-face grin.

It took Iran eleven years to win a verdict in the platform bombings case, and the United States repeatedly tried to get the case dismissed. Both sides presented dozens of experts and affidavits. One explanation for Iran's win is that the court wanted to set limits on what the United States was allowed to claim was self-defense in light of its 2003 invasion of Iraq. I ask why he thinks the court ruled in favor of Iran. "They weren't able to show that what they did was self-defense," he says. "Their actions did not amount to enough force to be self-defense." The attacks were *too* symbolic. It's interesting that Zeinoddin feels the case was won on the lack of strategy behind the attacks—in other words, the United States' muddled policy—rather than the very idea of U.S. dominance in the Gulf, and it shows how much a fixture the American presence has become.

At the end of the meeting, the PR lady hops up and says she needs to book the helicopter to take us to the platforms. Aresu squeals. And as our PR heroine almost skips out of the room, I notice that the back of her coat has an extra-tall slit, so she can take long strides.

### The Persian Gulf

Off to the Gulf. Early the next morning we arrive by plane on the island of Lavan, which looks like a strip of baked pie crust from the air. It's no more than a sandy oval with a refinery, three flaming gas flares, and some puffs of dark smoke. Temperature: somewhere over 100 degrees. Ambience: steam room. In our coats and scarves, Aresu and I are wilting fast. We are accompanied by a PR man from the offshore oil company, a serious man with a beard, glasses, and the tieless look of the faithful, who tells us that women are not usually allowed on the island. Our other escort is the head of offshore security for the oil company; he seems delighted to have us. Around us are bunkers, cut into the sand during the Iran-Iraq war.

A bubbly, boyish man in a blue jumpsuit runs toward us. Mr. Ebrahimi is in his late forties, with wisps of irrepressible chest hair curling out of the neck of his jumpsuit. He is going to give us a tour of one of the bombed platforms, named Salman. "I love Salman Complex," he says to me exuberantly in English, clutching his chest. Then he continues in Farsi. "I have spent more time there than with my family."

The Iran-Iraq war appears the moment we sit down to tea and small talk with Mr. Ebrahimi. His cell phone rings. He looks distressed, stares at the phone, shakes it a bit as if deciding whether to answer it, and then becomes very agitated as he talks to the person on the other end. His wife, Aresu whispers, seems to be unable to take their children to the first day of school and is asking that he come home immediately. As Mr. Ebrahimi continues negotiating, his voice wavering between outrage and pleading, the oil company's chief of security says that the war was bad for the company's workers but it was worse for their families, who stayed on shore and worried themselves sick. Mr. Ebrahimi hangs up but sits in his chair for a minute looking distraught. "It was hell for families," the security chief finishes flatly. "My wife still hasn't recovered from the stress of worrying about the war," Mr. Ebrahimi adds. "She has some kind of paranoia, she's very stressed and scared of many things. She's not able to take care of the children." He pauses for a moment and describes the intimate effect of the war as "a robbery."

After Iraq invaded Iran in 1980, Saddam began bombing Iran's oil facilities to deprive the country of the money it needed to buy arms and fight the invasion. Iraqi attacks on oil facilities reduced Iran's oil exports by more than half, and in 1986 the Iraqis attacked Iran's export terminals on Lavan, where we are now. In all, Iraq attacked Iran's refineries thirty-four times using dozens of missiles. Shortly after the war Iran's oil ministry said that the country lost $11 million in oil revenues every day for eight years. As the oil money shrank, the population had to make up the difference. By the mid-1980s Iran was rationing food so that it could spend half of its total revenue on importing weapons.

Oil workers became targets for Iraq and simultaneously responsible for keeping the country afloat financially. As much as the "human-wave attacks" of young Iranian soldiers throwing themselves at the Iraqi army were saving the country, the oil workers were also fighting. But they could not defend themselves. "It's not possible to make people safe on the facilities," remembers S. E. Jalilian, a former director at the national oil company, who says the company lost many workers to the eight years of bombardment. "Facilities are undefended. They're more like a home or a civilized zone. There's usually a specific procedure to produce oil, but during the war it wasn't possible. You just wanted to protect your production and try to refurbish as quickly as possible." Surprisingly, he becomes a bit nostalgic about the war. "During the war, something happened to the workers. People became more self-sufficient. They learned to protect themselves and survive. It was hard, but it was nice also."

**An hour later** we are strapped in a helicopter, wearing ear protectors and flotation vests, skimming across the surface of the Persian Gulf. Aresu is grinning madly and asking me to take her picture. I feel giddy.

Below us, the Gulf glistens as if it's made of grayish green jelly. The rippling waves form unexpected herringbone patterns, and sometimes the shadow of our helicopter appears below. The moisture in the air refracts sunlight so that we seem to be flying through illuminated cotton. A little fishing boat shaped like a curled slipper, or maybe a melon rind, appears and disappears below us.

Seen from space, the waters of the Persian Gulf look like a balloon, with the narrow neck facing east the Strait of Hormuz, through which all water, ships, and oil coming to and from the Gulf must pass. Iran dominates the entire north side of the Gulf, from the strait to the tip near Basra, Iraq's sole port. At the western tip of the balloon, next to Basra, lies Kuwait. Saudi Arabia dominates the southern side, on which Qatar appears as a small speck and Bahrain an even smaller speck. Closer to the strait sit the larger territories of the United Arab Emirates (UAE) and Oman.

The waters of the Gulf are said to be forty-seven times more polluted than the open sea, and the countries on its shores share the destiny of oil producers, but they don't feel a common sense of origin or purpose. They are a mixture of Arab and Persian, Shia and Sunni, monarchy and theocracy, and they live in a tense dynamic with each other. Their borders have been in dispute since the 1920s, when the British drew them according to their convenience. The UAE, for example, has territorial quarrels with Saudi Arabia, Iran, and Oman. A few other disputed borders: Iran-Iraq, Iraq-

Kuwait, Bahrain-Qatar, Qatar-Iran, Qatar–Saudi Arabia, Saudi Arabia–Oman, Iran-UAE. And Israel is always just over the horizon.

Gulf States deal with this perpetual insecurity by spending oil money on weapons. Between 1997 and 2004, UAE, with a population of just 2.5 million people, dropped $15 billion on arms (out of the whole region's $65.5 billion). Beneath where we fly, in other words, there are 3,395 missiles, 3 submarines, and 303 supersonic combat aircraft, as well as piles of tanks, helicopters, and armored cars, to mention those that were purchased just within the last eight years. Also below us is a U.S. naval base in Bahrain, any number of U.S. naval vessels, and perhaps two thousand marines floating around on amphibious vehicles.

An oil platform appears in the distance. Tiny boxes stick above the surface of the sea, reminding me of dental work—little bridges and crowns. As we get closer the boxes become platforms. A flare burns nearby. The helicopter banks and slowly approaches the top of one of the platforms. The sensation of landing on what appears to be a precariously small box in a large sea of undulating jelly makes me queasy. Over the door of the helicopter a sign reads, in red letters: DO NOT PANIC. We land on a platform rusted to lace on the edges. Through the filagree, the gray-green sea heaves. Out of the helicopter, Aresu and I clutch our head scarves and run doubled over to the edge. There is no handrail at the top of the stairs, which hang out over the water, so we drop to our knees to put our feet on the steps. In the steamy air, my scarf is sticky and my feet are blurry and far away.

Two careful flights down, I stand at a grid of twisted, rusting girders. Mr. Ebrahimi says this platform was bombed by Iraq during the war. Twenty years later, it is still not completely repaired. Farther down: more floors of twisted girders and two corrugated buildings, really glorified sheds, on the floor of the platform, which is still extracting oil. On the way back up, at the third level, two men in plastic sandals stand holding a tray of cold drinks in small plastic cups. Mine is sugared rosewater, which I sip contemplatively while staring at the enormous twisting melted girders, now velveted thickly with rust. Steel begins to deform at 2,800 degrees F. I'm trying to imagine the inferno of the platform burning, but there's ice in my cup, and the rosewater reminds me of the sachets my grandmother used to put in with her clothes.

We get back on the helicopter and take off into the cottony light, banking around a large platform called Reshadat, which the United States bombarded in the fall of 1987. It still has not been rebuilt, and an oil slick extends from its base. Something yellow also appears to be leaking from a pipe.

Now our helicopter is bearing down on Salman Complex, a ghostly collection of platforms, catwalks, satellite platforms, and flames in the morning's hot mist. There are twelve platforms associated with Salman, and yet, as we descend, it strikes me as a lonely place, lonelier than an island—more like a collection of solitary nests on sticks.

"Welcome to my home of twenty years," shouts Mr. Ebrahimi as we land. Below the helicopter pad, the catwalks are lined with flower boxes full of scraggly geraniums, as if this giant industrial complex were a cottage. A hundred thirty men live on Salman, and they all wear blue or red jumpsuits. We're led to a cold lounge postered with warnings about the dangers of heroin and hashish, while our hosts run off to put on the fresh jumpsuits they've apparently been carrying in their briefcases. Now Aresu and I are even odder-looking, hourglasses among the stick-men. Someone wonders if there have ever been women on Salman.

Mr. Ebrahimi returns, still seemingly overjoyed to be at his other home. "Would you like to see the oil? It's beautiful." We're off on a dash around the complex.

Salman is like one of those science fiction machines that is so complex that it becomes awkwardly alive. Eight main platforms, some of which are connected by catwalks over the sea, and under that forty-two valves; fifty wells; eleven drilling, production, and processing platforms; and a 150-mile pipeline to carry the oil back to land. We are whizzed from one end of the platform to the next, visiting the flares, a water-filtration room, complex mixtures of pumps and computer controls, separators, dehydrators, and turbines.

Salman is a gas-oil separation facility, which means that when the oil rises from the wells, the gas and water accompanying it are removed. The water is treated with chemicals and reinjected into the reservoir below, while the oil is piped away to the mainland. Among these big machines, with their fantastic complexity, the men are cheerfully submissive to the platform, as if they've married someone far larger and more powerful than themselves. Wherever we stand, there are pipes carrying oil somewhere else—up, down, sideways—always headed out of sight. In a pretzel of pipes on one of the upper platforms, Mr. Ebrahimi excitedly draws off a bucket of light, warm oil and dips his fingers in reverently, as though he's touching fresh milk. Then he holds them up and I, caught up in the moment, put out my finger to touch the oil. It is only a little warmer than the air.

In the 1970s, Salman produced 243,000 barrels a day, but as the reservoir depleted, the amount of water coming out of the wells increased until by 2000 it was producing mostly water and 80,000 barrels of oil a day. Falling production is a double problem—not only is there less oil, but the costs of disposing of the water rise, so the well's profitability falls dramatically. Since then the company has invested in the field, drilling horizontal wells and forcing water into the reservoir to push out around 130,000 barrels a day.

On the day we visit, oil prices have passed $50 a barrel for the first time ever because a young unknown Nigerian militant has declared war on his government. Thanks to him, and to all of the other pressures on the international oil supply system, the oil coming out of Salman is worth about $6.5 million today.

On one of the catwalks, Mr. Bayat, the security chief who has accompanied us from Lavan, points out a very distant platform that he says belongs to Abu Dhabi, part of the UAE, which owns a portion of the Salman field. Past the orange and yellow flames of the gas flare, I can make out a dim structure. "Sharing makes you closer, though it can also create problems. When the commercial goals are foremost in everyone's mind there are no problems, but when those are replaced with political goals it makes for war." He puts his hopes in these arrangements for the future, as insurance against war. "I think the politics have changed," he says. "Peace is good for business."

U.S. sanctions against Iran keep the engineers on Salman from getting parts and services for American-made equipment. "My life is spent with compressors," says Mr. Ebrahimi, explaining that he and his colleagues have reverse-engineered the compressors to build parts for them and fix them. But too much time is spent fixing things that would be better replaced. "When they are out of service the oil production decreases. The production problems are quite different than the political problems." The engineers go on to describe attempts to fix American equipment without either parts or advice from the manufacturers. They've tried e-mailing the firms, but the companies look at the equipment serial numbers and refuse to reply. After the Bam earthquake disaster in December 2003, there was a brief break, when sanctions were lifted. In the control room, the engineers ask to have their picture taken in front of a bank of computers and a plastic plant before going into details about their frantic e-mails to equipment manufacturers during the disaster. But soon that window closed, and they were back to making do.

**The afternoon** is wearing on and the helicopter will soon return for us, so Mr. Ebrahimi can no longer put off his story.

On April 18, 1988, Mr. Ebrahimi awoke—by that time he was used to sleeping in his boots and life jacket—to the news that three American warships had surrounded the platform, but he went to work on the air-conditioning system, as usual. On the lower levels of the platform, he says, there were Iranian military to protect the facility from attacks by Iraqi planes. U.S. accounts say that the platforms were used to launch attacks against ships.

At 7:30 A.M. the Americans demanded that everyone leave the platform within thirty minutes. Mr. Ebrahimi remembers that the manager felt there wasn't time to evacuate the entire platform. "We better ask them what's going on," he said. But the United States refused to negotiate, watching the "frantic" activity on the platform from a distance. At 8:04 they began shelling the platform. Mr. Ebrahimi says the United States blew up the platform's lifeboats; American accounts say the military installation was the first to get shot. An Iranian tugboat radioed for permission to pick up "a large crowd of converted martyrs gathered at the northern end," and the United States stopped firing for a moment as the men ran from the platform.

Mr. Ebrahimi managed to reach the tug, but one of his colleagues spent an hour and fifteen minutes in the water without a life jacket. One was found swimming completely naked. All were scared of sharks. "Some didn't dare jump. They went into the shelter and lay down." The United States continued attacking platforms while a few Iranian ships tried to pick up the men in the water. Then a U.S. helicopter dropped marines onto the platform, where they set so many explosives that the concussion from the blast could be felt five miles away.

Mr. Ebrahimi starts to move his hands around his face as if he's swatting mosquitoes. "It was the worst psychological situation. None of us were trained in the military. We couldn't defend. I myself was in the middle of the sea, with no gun. It upsets you; it makes you full of hatred." He is twisting a tissue, and sweating under the air-conditioning.

"And then there was Salman blowing up. That was the worst emotion I've felt, ever. That was my home and it was exploding. I didn't know if my coworkers were in there or not." By the end of this story he has become a different person: angry, moving his body to a different rhythm from the boyish man who welcomed us.

He spent the rest of the day on the tug watching the smoke of the burning platform. "Even after four or five hours, we couldn't get enough of watching the smoke of where we used to live. When you see your colleagues jumping, the situation makes you crazy and full of hatred. I still get woken up with shocks from it."

That hatred and frustration, the continuing shock from the long-over war, is a primal influence for the men who are coming to power in Iran now. "For the new generation of conservatives, it is the war with Iraq, not the revolution, that is their defining experience," Ray Takeyh testified in Congress. "Their isolation from the United States, suspicion of the international community that tolerated Iraq's chemical weapons use against Iran, and their continued attachment to the tenets of the revolution define their ideology."

During the Iran-Iraq war, Iran lived surrounded by enemies. The country's efforts to foment Islamic revolutions among the Shia populations of neighboring Iraq, Saudi Arabia, and Bahrain led other states to support Iraq. Saudi Arabia gave Iraq at least $16 billion in loans and arranged for arms shipments through Egypt. The United States officially maintained that it was neutral, but it was providing so much assistance to the Iraqis that it was really involved in an undeclared war.

April 18 ended with a clear picture of the implications of U.S. assistance. While the Americans were bombing and battling the Iranians in the Gulf that day, Iraq was using chemical weapons on Iranian troops on the Fao Peninsula. There's no direct evidence that the United States helped Iraq plan the chemical attack, but the CIA was providing Iraq with satellite photos of Iran's positions. Interviewed by the *New York Times* in 2002, Colonel Walter P. Lang, a senior defense intelligence officer in the 1980s, said that both the CIA and the Defense Intelligence Agency were "desperate to make sure that Iraq did not lose to Iran. The use of gas on the battlefield by the Iraqis was not a matter of deep strategic concern," he said, compared to the fear that Iran would attack Kuwait and Saudi Arabia.

"Time is not on our side anymore," the Speaker of Iran's Parliament, Hashemi Rafsanjani, reportedly said on Tehran Radio after the attacks. "The world—I mean the anti-Islamic arrogant powers—has decided to make a serious effort to save Saddam and tie our hands."

The war with Iraq, the U.S. involvement, and all of Iran's bitter isolation are still alive in here. While I'm here, editorials in the paper debate the impact of the war, and one dredges up an old Khomeini quote—"Iran showed that the global powers are deceptive"—for the edification of the people too young to remember the war. Whether at the massive cemetery near Khomeini's tomb, or around dinner tables, the war is frequently an undertone in Iranian conversations. What Iranians learned was that the country could not rely on friends, institutions, religion, or money for protection. Their only support was their oil money and their ability to be enough of a military threat to deter attacks.

**As the helicopter** comes in to land on Sirri Island, a small stretch of roasted sand and palm trees, I regret that I don't know more about how to identify missile installations from the air. Sirri is thought to be studded with Chinese-made Silkworms, designed for defensive attacks on ships. What I see below me looks more like an inhospitable picnic area in a state park, but the missiles are said to be aimed at the Strait of Hormuz.

Every day, nearly 16 million barrels of oil pass through the narrow, two-mile-wide outbound shipping lane of the strait. That's 40 percent of the world's traded oil, and, on a daily basis, a third of the oil western Europe uses and three-quarters of what Japan imports. Close the strait and the world's economy closes too.

The missiles, along with three submarines, an air force, and an array of smaller boats, give Iran the ability to take the strait (and the whole Gulf's oil) hostage in the event the country is attacked or feels threatened. "We have told the Europeans very clearly that if any country wants to deal with Iran in an illogical and arrogant way . . . we will block the Strait of Hormuz," said a spokesman for Iran's nuclear agency.

**The manager** of Sirri's oil facilities is Gh. Aslani. Around fifty, Aslani is more courtly than imposing, with a thin rim of hair and a grizzled chin. He invites us to sit at a long table in his office, where one wall is rock and the other is wood paneling. It feels like a rec room in a ranch house. On the table are tissues, Cokes, and a little stack of macaroons that remind me of the tank farm outside. Aslani was present when Iraq bombed the first refinery in 1980, and then moved from one target to the next during the eight-year war. Educated in the United States, he begins to talk about the refinery on Kharg Island, which Iraq bombed almost daily when he worked there in 1986.

"The red alarm went off one night and we went to the shelters for an hour and a half," he starts calmly, "and then a bomb hit a storage facility and exploded. A pipeline nearby also got hit so we came out of the shelter and saw the huge fire. They said the fire was getting close to more storage tanks, and they were worried that the heat was being transferred from one tank to another, so they tried to cool the tank with water. But there was another gas pipeline and there was oil leaking from a manifold. The gas line exploded. I was twenty meters away and forty or fifty people got injured. Everyone was trying to escape with broken legs and hands. The fire grew to one hundred meters long and vacuumed people into it. They couldn't run."

He tells his stories with precise intensity, as if mapping out an engineer's diagram of Guernica. The night of this one fire he stayed up spraying water on the tanks, trying to prevent another explosion. He was burned twice. One-fifth of the men of his generation were killed in the war.

Aslani was working on Lavan when the United States attacked the platforms. He saw the attacks from the oil company's helicopter and was afraid to return. He becomes more angry as he talks, as if he's reliving times not far removed from his daily life. "What do you expect? We were unarmed people."

By the time Aslani has finished, he is furious, wound up, his eyes like those of an angry bird. He rails at me. Aresu hits me under the table to end the interview. But every time I try, he begins again. "I believe the U.S. government and the people are different, but sometimes the government magics the people," he says.

More anger. He becomes very confrontational, ranging from the personal to the impersonal. The fiction of the government and the people being different is easier to keep when you're thousands of miles apart. Aresu hits me. She writes "Let's go" in my notebook. The Americans who boarded Salman's platform wrote obscenities on it, Aslani says. His sparse hair stands in a furious haze above his scalp. Why do the Americans act the way they do? he wants to know. "You know what I mean. We like freedom and democracy, but we need to build it ourselves."

Inside Iran, Khomeini called the war a "blessing," because it provided an excuse for the most radical elements of the government to make it more authoritarian. The U.S. attacks were aimed at the Iranian decision makers, but they found their real target in people who were ambivalent about the regime.

One Iranian explained his reaction to the platform attacks this way: "We didn't like the U.S. intervention in 1988 because it was hostile. The government then used this hostility in propaganda, which allowed them to forget their own crimes." But then he goes on to explain the second disappointment of the attacks: "The U.S. is a superpower, like the father of the family, and we expect a better understanding of cultures."

There was another issue that no one mentioned, and that was that the United States targeted the oil company workers and the Iranian navy when the group that was really laying the mines was the Revolutionary Guards, who saw themselves as martyrs-in-waiting. Cheap but effective, the Guards habitually attacked tankers from tiny rubber boats armed with machine guns and rocket-propelled grenades. They did not answer to the Iranian navy, which had been buzzing around the Gulf furiously trying to remove the mines the Guards had laid. It's not clear whether the United States failed to recognize the distinction between the Revolutionary Guards and the navy and oil workers, or whether decision makers deliberately ignored the difference.

So it's not surprising that Aslani thinks that Americans have no idea what they're doing with their "decisive military actions" and their "vivid pictures of right versus wrong." Now he wants to know why the American people are not aware of the kind of chaos and long-term war they started in Iraq. He wants to know if we have any idea what's at stake. What I didn't realize at the time of our conversation is that Aslani and his facilities are still a military target. Israel has threatened to attack Kharg Island and other oil depots to put an end to Iran's nuclear program.

Aslani becomes angry again. The rocks on the wall behind him seem to be moving. Sparks dance in the periphery of my vision. He asks again why Americans don't care. I say we're self-absorbed. He doesn't like that answer. Aresu is poking me. But what, really, can I offer?

When someone outside Iran talks abstractly of how U.S. military hegemony supports the flow of oil to world markets, it's a dry observation. When an oil analyst postulates that America's willingness to use military force was a factor in creating cheap gasoline in the 1990s, it seems remote. But sitting across the table from Aslani, the personal recipient of some of these military "messages," it's quite immediate.

Operation Praying Mantis was judged a success by American policy makers. Former CIA analyst and National Security Council member Kenneth Pollack refers frequently to the operation in his book *The Persian Puzzle* as proof that, "given a clear demonstration of America's willingness to employ overwhelming conventional power, Iran would back down." The authors of the military concept of "shock and awe"—a theatrical display of force to overwhelm and confuse the enemy—refer back to the platform attacks as proof of its effectiveness.

But did the United States draw the wrong conclusions? It more or less won the one-day war in 1988, but Iran has been providing covert funding for terrorist groups for nearly a quarter century. And it now dominates the Strait of Hormuz in a way it could not in 1988. In a 2002 U.S. navy war game, a defending force (presumably Iran) commanded by a former marine corps three-star general and consisting of nothing more than small boats and some aircraft attacked the simulated U.S. Navy Task Force entering the Gulf and sank most of it. According to the ex-general, the navy "rigged" the games, refloated the American boats, and disallowed his tactics to show that the United States could win such a battle.

When MIT professor Barry Posen looks at Operation Praying Mantis, he concludes that it was expensive, perhaps too expensive to be repeated. Fighting in the shallow waters near oil facilities puts the United States at a distinct disadvantage by pitting millions of dollars' worth of ships against small but lethal local fighters. The mine that the USS *Roberts* hit in 1988, for example, cost just $1,500, and was made in the Soviet Union from World War I–era plans. The hole it made in the *Roberts* was twenty-five feet across. "The question that needs to be asked is: Is the cost of the work-around cheaper than a fight?" says Posen. "You're fighting against people who live there, with cheaper weapons. They can make it very expensive. A real brawl. It's worth thinking about what you give and what you get."

Another problem, says Posen, is that using military strikes makes the idea of them less potent. And as the war in Iraq drags on, it's clear the United States has lost its mystique of overwhelming power. "Sometimes to have control you have to not assert it. You have to be a gray eminence."

For me, the abiding lesson of Operation Praying Mantis is Aslani's anger. Winning, particularly in the politics of the petrostate, is little more than the start of a long war. A military win in an oil field is not like a military win on a battlefield because everyone has to cooperate to get the oil out of the ground and sell it. In 2006 President Bush promised to "make our dependence on Middle Eastern oil a thing of the past" by 2025. Even if that is possible, which I doubt, we still have nineteen years of living together left.

By the end of the discussion, Aslani is speaking agitatedly in Farsi while gesturing at me with his eyebrows. I ask Aresu what's going on. She says he'd like to give me an English version of the Koran, but his only extra copy is in French. I take this as a sort of housewarming gift, a "welcome to the Gulf neighborhood and please don't be a bore" statement rather than proselytizing. After I explain that I already have an English copy at home, we're finally able to leave his office.

It's after dusk when we exit Aslani's office to walk through the foggy refineries and loading docks of Sirri. The pipes and tanks are beautiful in a ghostly, austere way. The air is hot and tired, as if it's been breathed twice already by someone else.

Sirri Island is one place where oil could have mended the differences between the United States and Iran. In 1995 Iranian moderates awarded a $500 million oil contract to ConocoPhillips to invest in the facility. The company had to abandon the project when President Bill Clinton signed further sanctions against Iran later that year, because of Iranian officials' continued support of the fatwa on Salman Rushdie and possible hand in terror actions in Lebanon, Germany, and Saudi Arabia. The French company Total took over the project, and those in Iran who hoped for a mediating U.S. presence were out of luck.

That night Aresu and I stay in an empty guest apartment on Sirri. We are not just the only women, we are the only ones ever, which makes us freaks. Several men come to the door to ask if we need anything and look us up and down. Aresu slams the door, rips off her head scarf and coat, and flings them in a sweaty pile. She turns on the TV. Oprah is on, and she starts mimicking her questions. Aresu is a wicked mimic. And she is also furious. Furious, I think, at ten things at once, two of which are the United States and me, and some are other things that I don't know anything about and perhaps never will. She's sitting in front of the TV simmering. The more time I spend here, the less I understand. If it wasn't Iran, and we weren't on this bizarre militarized island near the world's most important oil chokepoint, we'd go out for a walk or to a coffee shop, and maybe we'd both learn something. But this is Iran, and we have to deal with things as they are: It's bedtime.

**In Tehran,** two government economists sit under a map of the world's oil basins. The map takes up the whole wall, and in the center of it the Middle East is a thick dark whorl, dwarfing all other reservoirs, creating a spot so dense it has its own gravity. Fifty-seven percent of the world's oil reserves. Forty-five percent of the world's natural gas reserves.

"We are always analyzing the relationship between energy, the economy, and diplomacy," the older economist says. In front of him, a sign in English reads: THE BUCK STOPS HERE.

"If you look at the oil reserves in the world, Norway, Mexico, the U.S., and even Canada will be exhausted in the next ten years. And if you look at the outlook of 2020, the U.S. will demand twice as much imported oil as they do now. We know most of that oil will be in the Middle East. I think it was Kissinger who said that oil is the blood of the American economy."

It is perfectly clear to them, and to other Iranians, that the United States is not leaving Middle Eastern oil or waters anytime soon. "What you're seeing in Iraq is part of a power struggle over who will control the flow of resources in the region," says Bijan Khajehpour, a prominent Iranian oil consultant. "The big question is: Can the U.S. live with an anti-American Iran or not?"

Iran, like Venezuela and other OPEC suppliers, is trying to decide how to get the most, strategically and economically, out of its oil. "There are three ideas about what to do with the oil here. One of them is to really increase output to become a powerful swing producer like Saudi Arabia; another is to increase to exporting 5 million barrels a day. The third thinks that they're making a mistake by selling oil now. Wait until it's $100. Just sit on it," says Khajehpour.

The depletion of Iran's oil fields and the need for new (foreign) investment is high in the minds of the economists. They say that it takes about $7,000 in investment to get an extra barrel of oil per day from a new oil field in Iran, but it takes $14,000 investment to prevent decline in an old one. Old fields like Salman are getting expensive to operate. In 2005 the oil minister mentioned that Iran's fields are losing 400,000 barrels of production a year, and the investments the country needs to maintain its output are huge.

Why, the economists would like to know, doesn't the United States invest? One leans toward me. "Let me put it this way. Twenty years of sanctions. What did they gain? Our capacity is 4 million barrels a day. We'd like to produce 6 million."

The other economist interrupts. "If we had thousands of companies here, we could push the government to be good. You could stimulate the Iranian economy and forget the sanctions and foster private industry. If there are powerful companies in Iran, they'll push for more power, to moderate the discussion. But if Iran has nothing to lose, it can keep up the hostile activities."

If the sanctions were supposed to make Iran an outcast, they failed. Through the last ten years, Iran's stature and influence in the region have grown. Its years of building business relationships in the Gulf have paid off. "The U.S. influence in the Gulf is waning while Iran is ascendant," says Rachael Bronson, Middle East expert at the Council on Foreign Relations. "Iran is clearly exploiting opportunities and America stumbles from Iraq to Afghanistan to Lebanon. Iran's come out well."

And, with the last war in mind, Iran is using its oil and gas to protect itself. In late 2004 it agreed to sell China natural gas and oil for the next twenty-five years, a deal that may be worth as much as $200 billion. In addition to money and an auto assembly plant, Iran has tried to join the Shanghai Cooperation Organization, a nascent alter-NATO that includes Russia, China, Tajikistan, Kazakhstan, Kyrgystan, and Uzbekistan in an effort to be a counterweight to "U.S. power unchained." Iran sees itself as a "front-line state" against U.S. hegemony. Its back line, it hopes, is China and Russia, which recently agreed to sell Iran $1 billion worth of defensive missiles.

Suddenly sanctions are not only keeping American companies out of Iran, they are also, in effect, providing China with lower-cost access to the world's largest reserves by eliminating U.S. competition. And they are making Iran less vulnerable to diplomatic and military pressure. It was such a perfectly bad deal for the United States that I half expect someone to tell me the United States has engineered the whole thing as part of some larger conspiracy.

When Washington threatened Chinese companies that do business in Iran with sanctions, China scoffed. "We don't care about the sanctions and I don't think the Chinese government cares," China's commercial attaché in Iran told *Fortune*. "Why should we ask America's permission to do business? We are a sovereign country."

It is, in a way, a replay of the 1940s when Americans entered a relationship with Iran and were welcomed because they were less intrusive than the colonial British. "If I may say," says the older economist, "the U.S. had a beautiful picture after World War II. For two hundred years we'd pictured Europe as an exploiter. But with these interventions they ruined it. We hate the double standards—one for Israel and one for Iran. People don't believe the U.S. supports human rights because the speeches are different than the reality." Now it's China's turn to have the beautiful picture.

Like Venezuela, Iran is on a petrodiplomacy binge. The country is in talks to build a $4 billion gas pipeline through Pakistan to India. And while this so-called Peace Pipeline could strengthen ties in a tough neighborhood—Iran and Pakistan are often at odds, as are Pakistan and India—the United States has vehemently opposed the deal because it involves Iran. India's oil minister bristled when asked about U.S. opposition in 2005. "We will not come under the pressure of no country and a decision will be made in the interest of our people," said the Indian oil minister, adding that Pakistan agreed.

The economists were thinking about other possibilities. Iran also has the world's second largest reservoir of natural gas. "If we do a gas pipeline with Turkey, we're connecting Iran to Europe," says one. There was talk of building a pipeline to connect Iran to Russia's gas pipeline network, somehow allowing Iran to get in on Russia's monopoly on gas supplies to eastern Europe. Even without its nuclear weapons program, Iran has lots of cards to play. But the United States, which has dealt with the country mostly through sanctions and threats of military action, has fewer and fewer options and a need for Gulf oil that grows every day. It's easy to see why the fantasy of regime change has so many fans in Washington—working with Iran, working through the issues of the Gulf, and building relationships is going to be hard work.

**On our last** day in the Gulf, Aresu and I visit the Nasr platform, which Anthony Rodriguez's ship, the *Wainwright*, bombed in 1988. The platform has been almost entirely rebuilt, and at its base we can walk on pathways just a foot above the greasy sea. Under the platform I see strange blue and yellow shapes. Gradually they assemble themselves into angelfish or some cousin of them. Only these are the size of an open newspaper. And within a minute there are hundreds snooting the water and flashing their flat bodies. The men throw them bread and an angelfish melee begins: fins, tails, water bubbles, and soggy bread. As they surface and retreat, I barely make out their bodies by rearranging the blue and yellow puzzle pieces that appear on the surface. I see something that looks long and thin, like Rodriguez's sea snakes, but it turns out to be a trick of the light against the striped bodies of young angelfish. For the first time, I see the Gulf outside of its commercial and strategic context—as a place for fish.

It's tempting to dream that if the Gulf were just a big aquarium, all these tensions would go away. "As long as we define the Persian Gulf region just as an energy resource," said Mohammad Khatami, Iran's reformist president through 2005, "the destructive presence of alien forces would continue and improper domestic rivalries would overshadow peace and stability in the region." I find Khatami's vision plaintive, but it's hard to imagine a time when the Gulf will not be defined by its enormous energy holdings. And it is futile. Better, I think, to stop seeing the

oil as solely a cause of tension and start looking to it as a route to a solution. Just as oil creates rivalries in the Gulf, it also draws countries
together. Oil exporters and importers have overwhelming common interests, no matter what their other differences. With work, perhaps this era
of conspiracy theories can give way to a time when the countries of the
Gulf feel they control their own destinies.

Aresu and I run around on the pathways under the oil platform, tossing bread and trying to take pictures of the fish. Whatever we don't get
about each other—and there is much—we share a common obsession
with oil, and a sense of the absurd. We end the trip with a dogged search
for stuffed sharks, which we find at the last possible minute, less than half
an hour before our plane departs. I buy a glassy-eyed baby, its smile made
of little serrated teeth. Aresu, of course, buys a much larger shark.

# 10  NIGERIA  *CALLING THE WARLORD'S CELL PHONE*

His Royal Highness J. C. Egba steps off the road into the damp grass in his shiny dress shoes and begins walking unsteadily toward Nigeria's first oil well, named Oloibiri Number One. Wearing long chartreuse robes, strings of coral beads, and a small black hat with a stylized image of a crown beaded into it, His Royal Highness, a traditional king from the nearby village of Oloibiri, lurches toward the rusty pipe in the corner of the field, narrating the story of the day Shell engineers found oil in a test well. "The oil rained down on the swamps and fishes," he is saying. He is holding a scepter with a tuft of white hair at varying distances from his body the way a tightrope walker uses a stick for balance. I am following him, worrying about the elderly king falling in the slippery grass. Thick dark clouds have lined up above us.

The well now sits within a rusty fence at the corner of the field, abandoned since the oil ran out in 1977. In 1956 this leafy spot was the very beginning of Nigeria's oil industry. Now, on good days, Nigeria exports 2.5 billion barrels, providing 80 percent of the national budget and making up 10 percent of U.S. oil imports. Under the Niger Delta—an England-size patch of mangroves and dizzy curving creeks—sit another 36 billion barrels of oil. Nigeria hopes to double its output by 2010. But on bad days, when violence hits the delta, exports fall, and the price of oil rises around the world, making this deceptively bucolic patch of wetlands central to the industrialized world's purse, if not its well-being. The chartreuse king reaches out his scepter-free hand to grab the fence around the well.

From behind us there is a loud voice. "Come. Out. Here!" The voice has a panicky tremor. "There may be problems soon!" We turn to find a small man in a fury. He seems to have emerged from the wall of foliage, but he is wearing a pale green jersey with a ballpoint pen incongruously clipped in the pocket, as if he works in a suburban auto parts store. He is vibrating with rage. He moves toward us again, shouting "It's very unfortunate that a chief is trespassing." I search the edge of the clearing again for houses or other evidence of ownership, but I see only green. HRH shakes his stick and says he's the king. He is not trespassing.

"You say you're not trespassing? I will have the police arrest you right now!" The man's voice has become even louder, a large volume of air rushing through a constricted place in his throat. His nose has developed an extra wrinkle of indignation running at right angles to the bridge.

"How old do you think I am? I am fifty years," the man shouts. "In 1956 I was very small. I lived over there with my mother!" He waves into the green behind him.

Some men are riding by on bicycles. The furious man waves them over to help arrest us. I am taking notes, and thinking that just because this is inscrutable doesn't mean it won't escalate. Nigeria defines the worst environmental and social disintegration that oil production brings. With a thousand battle-related deaths a year, the Niger Delta has the same level of conflict as Chechnya. Only six months earlier, in the fall of 2004, the delta was a battleground where thousands of men in two large militias fought each other and the government.

Though I'm in the delta to see the relationship between oil and conflict, today was supposed to be a day off. Dennis, who works for a Nigerian nonprofit called Our Niger Delta, suggested that it might be interesting to pick up the king and head out into the hinterlands, which locals call "the creeks." We drove to the village of Oloibiri to see where the oil industry started. But you can't escape conflict here. The young men move toward us uncertainly, their eyes shifting back and forth between the spectacle of the chartreuse king and me, the white lady.

Dennis tries to soothe the man by talking to him. HRH seems to have retreated deep inside himself, perhaps to another time. His eyes are a bit milky, and for most of the ride he has been happy to circle in reverie about the time in 1945 when he was the headmaster of a school and a future lieutenant governor was his houseboy. I don't blame him for preferring 1945.

"Oloibiri is an oppression name!" the man says, spitting. "I know you are King J. C. Egba! J. C. Egba wants to cause problems between Oloibiri and my village, Obake."

We are in the original trouble spot, Dennis explains. When Shell, or other oil companies, extract oil in the delta, the money goes to the Nigerian federal government. Shell compensates local communities for oil production, paying the most to "host" communities that claim the wells. Here all of the compensation money, or at least the bulk of it, went to Oloibiri, while Obake village got little, even though it was closer to the well. The doling out of the compensation money turns village against village all across the delta. Locals say Shell's strategy has been "divide and rule," and we could discuss whether that was the company's cynical intention or an aspect of its cluelessness. But we need to leave. And quickly. Apologizing, we hustle the king and his scepter into the car and drive away from the man who'd been staring at that rusty pipe, seething, since 1956.

As we pull away he shouts, "If you do not live according to God you will not live like a lizard." At least, that's what my notes say. After the fact, I do not think the furious man was crazy. He seemed relentlessly rational. In Nigeria, violence and oil have combined to make strife that consumers care about in a lackadaisical way. Crisis after crisis here has caused spikes in the price of crude oil—like distress signals that are getting louder and louder—that the industrial world pays for almost daily at the gas pump. But the violence is understood as ethnic (i.e., crazy and unsolvable), and when it threatens the U.S. oil supply, it's described as terrorism. During a month in the delta I saw an increasingly reductionist and violent approach to a monstrously unfair and complicated situation.

Oil in the delta is the cause of the conflict, the prize of the battles, the provider of the weapons, and, increasingly, the medium through which people send messages to the outside world. The violence here is complex and horrible, but it is, in a dreadful way, all too reasonable. "What's illogical about trouble in a community that has no light, no water, and no work?" says delta activist Dimieari Von Kemedi. "Whether it's your problem or not, one day it will be visited upon you."

**Hornitos in Venezuela,** Ngalaba in Chad, and now Obake and Oloibiri: These places sat silently by as the oil flowed out of their communities for much of the last century. The village has always been an internal problem for oil-producing countries to solve as they wished, while the industrial world's continuing access to oil has appeared to lie in tech-

nology, OPEC diplomacy, the NYMEX, and eventually the threat of sanctions or military action. Recently, however, cell phones, the Internet,
and CNN have made the rest of the world aware of the village's burdens.
More significantly, the global trade in both stolen oil and guns has made
the village more powerful.

This became clear in September 2004, when Alhaji Dokubo Asari, a
Nigerian militia leader, sent the price of crude oil at the NYMEX over $50
by threatening to attack his government and the country's oil companies.
Asari was one of many warlords in the delta, and this was just another
skirmish in years of ongoing violence. But Asari was well spoken ("Operation Locust Feast" was the name of his campaign against the government) and well armed (three thousand fighters with a wide array of
weapons). And he was well funded because his men were involved in
bunkering, or stealing crude directly from pipelines, which has turned
the delta's oil infrastructure into a perpetual font of money and guns.
Bunkering draws the militias, the navy, the police, the highest levels of
government, and international black market networks into an enormous
tangle that conspires to steal between 10 and 35 percent of the oil produced by Nigeria's fields. An opportunist, Asari cast himself as a self-
styled Robin Hood. "I don't engage in bunkering," he told Human Rights
Watch. "I take that which belongs to me. It is not theft. The oil belongs
to our people."

Contacting Asari was easy: Another reporter gave me his five cell
phone numbers. I flew to Lagos, Nigeria's megacity, and then went south
to Port Harcourt, the biggest city in the Niger Delta. Then I called one
of the numbers. Someone answered. He said he was Asari, and he
sounded amused by my call. He asked a lot of detailed but tangential
questions. His voice was smoky, like a DJ on a smooth jazz station. He
said to call back. I called the same phone later. A different voice answered. This man's voice was deep, a scratchy rat-a-tat-tat. He spoke
quickly and then fell so silent I could hear his breath, which clacked in
the receiver. This man was really Asari. He said to drop by his house at
seven on Friday morning.

It was too emblematic of my time in Nigeria that I couldn't tell the
real Asari from the pretender, but really, he could have been anyone,
from any village. Asari is the latest addition to the petrostate triangle:
the oil-funded warlord. And these warlords, with or without ideological

agendas, are starting to change not only the price of oil in the market, but also how consumers assess its costs, both in money and in human lives.

Similar arrangements are cropping up all over the world. In Colombia, right-wing paramilitaries are reportedly stealing seven thousand barrels of refined gasoline a day from pipelines. In the Strait of Malacca, off the coast of Somalia, and near the Iraqi port of Basra, gangs steal whole tankers of oil, either by offloading it to another tanker or by stealing the entire vessel and changing its name. In Chechnya, groups both for and against Moscow are refining siphoned crude in backyard refineries called samovars and selling the homemade gasoline. Reportedly, half the area's oil is being hijacked, and the Russian military is in on the deals. And in Iraq, an "oil-smuggling mafia" is allegedly sending its profits to militia groups that are fighting the United States and each other.

If large global or even regional markets for stolen oil became well established, bunkering cartels could become shadow powers, perhaps more powerful than drug cartels. One sign that the networks in Nigeria are both international and connected to powerful people came when Nigerian police intercepted a Russian tanker full of bunkered oil off the coast. The tanker was later stolen from police custody, and two Nigerian admirals were charged with the theft. "The amazing thing about the trade in illegal resources is that you see the same people again—in South Africa and in Iraq," says an investigator who prefers not to be named. "Some do diamonds and some do oil. The Mr. Bigs recycle themselves."

In Nigeria, Shell already has one eye on the clock. A report the company commissioned in 2003 said that Shell's "business life expectancy" was threatened by a "lucrative political economy of war," and it would be "surprising" if the company was able to continue pumping oil onshore beyond 2008 and still maintain its business principles.

All of this was on my mind while I waited to see Asari. When Dennis suggested visiting the first oil village, I didn't expect much, because the oil was long gone.

**Back in the** car with His Royal Highness, we are approaching Oloibiri village. The road peters out into a dirt ledge leading to a sinkhole with a few scarred homes in the distance. Oloibiri was accessible only by expensive boat rides until recently, when Shell agreed to build a road. But the road doesn't extend all the way into the village. "It's nothing to write about," the king says.

We get out and walk. The mud is a bluish mosaic of crushed snail shells. The first houses appear to have been built by people who were shipwrecked and used driftwood to make temporary shelters in a hostile land. Some unused power lines sag overhead. I don't know what the village of Obake is jealous of.

HRH starts his narration again as he leads us toward the muddy creek bank that laps near the village. "On the thirteenth of January 1952, the Shell people came up the river in their houseboats." The powerful boats caused big waves in the river, swamping the local dugout canoes. About ten people died, he says, when the waves caught their ropes and sank the boats.

The village is not happy to see HRH. A woman scrubbing something looks pained. Young men back away. A relative of the king's who traveled with us begins to talk rapidly and soothingly with some of the elders.

"The traditional ruler is a hotcake!" says a Nigerian conflict negotiator who's worked here. "Every traditional ruler automatically becomes a Shell contractor, receiving payments on behalf of the village. When the workers first came, Oloibiri was happy to be chosen as the host village to get the proceeds from the wells. Production ended and the village feels the king collected money for himself. Maybe he did, but not what they are accusing him of taking. They're also angry with him for not putting more pressure on Shell to develop the town." Although corruption offered some kind of democracy of distribution in other oil countries, in Nigeria the corruption didn't trickle down. A 2004 World Bank study estimates that 80 percent of the country's oil money goes to 1 percent of the population.

Now that the oil wells are dry, the negotiator says, Oloibiri has no way to pressure Shell for roads or clean water or electricity or any of the things that have been promised over the years. "There are no helicopters. There is no person to grab," he says, referring to the hostage-taking that has become almost the norm for frustrated villages in the delta.

And so the village's understandably sour feelings have settled on the frail frame of HRH J. C. Egba, which explains why he lives far away in the city of Port Harcourt. When Dennis first brought me to meet him, he was sitting in a gloomy water-stained concrete room, receiving visitors on five overstuffed armchairs with lace antimacassars. The scene struck me as Dickensian and tragic. I'm told that Shell still considers him the local ruler, but the residents of Oloibiri no longer accept him. More than oil, the village has lost continuity with the traditions of the past and, with that, faith in the future. "One of the effects of the oil is that people begin to see themselves in economic terms," says Patterson Ogun, a longtime area activist. "It breaks the cords that tie people and communities together."

At Oloibiri's muddy waterfront, the king continues the story of Shell's arrival. "After the people died, the community said these Shell people should go, but I let them stay down at the bend of the river. And then another report came that the crews were going after the village girls. I asked the crew if they were chasing married women. They said only girls. So I said okay, they're human beings." I turn to see if there's a negative reaction from the little crowd around us and see mostly ears, because people have averted their faces and leaned in, ears first, to hear the conversation.

"On the tenth of June, oil was struck. They had no control. It was coming up like bump, bump. Like a bomb. It drained to the river and the fishes perished." That afternoon the town and the crew played soccer. "We thought they'd bring us drinking water and electricity."

Nearly fifty years later, the village has neither. A village leader shows me the water system, one of two built by oil companies. Neither works. He shows me the hospital that Shell built—now a mossy heap. No patient was ever treated there. The schoolhouse, built by Shell in 1992, is closed and rotting. The delta is covered with hundreds of these useless and abandoned development projects, donated by Shell and other oil companies. Some of the projects were inappropriate, or so riddled by corruption that they never worked. Others arrived in villages with no supporting infrastructure—a hospital with no staff, schools with no teachers. Five years ago the American ambassador visited and wept, the elders say, but they're still waiting for results.

"Millions and billions of dollars have left here," one elder tells me. "It's like a snail. They've taken the flesh and left the shell." More than $300 billion in oil money has left the delta.

The delta has a very old system of land ownership, but over several decades of military rule, Nigeria's federal government laid claim to the oil under the land as well as the land itself, in effect colonizing the people who live on it. "It's easy to say it's oil," says conflict negotiator Onyoha Austin, who points to complex issues around land ownership and community relationships as underlying problems. "Oil has become the determining factor in Nigeria because the government is only about how to split up the oil money." Nigeria's minorities have struggled for power ever since colonial Britain set up a federal government that split power among three main ethnic groups: Yoruba, Hausa, and Ibo. The Ijaw, an ethnic group with 14 million people living in the delta, are a macrominority, too small to attain power at the federal level, too big to be ignored. In 1966 an Oloibiri native named Isaac Boro led a twelve-day revolution, voiding all contracts with the oil companies and declaring that the delta was an independent Ijaw ethnic republic. The revolution had hardly gotten started when the Nigerian army showed up in boats provided by Shell and blew the revolutionaries away.

In the late 1960s an Ibo group (the Ibo are the smallest of the majorities) began a movement to form a separate country that included the oil-rich delta. The brutal three-year Biafran War followed, fought in villages all across the delta. In the 1990s businessman and playwright Ken Saro-Wiwa led a peaceful drive for environmental and human rights for the Ogoni, a small minority group living not far from Oliobiri. He was executed by the military government of Sani Abacha in 1995.

In 1998 a group of five thousand young Ijaw were back in Boro's home village of Kiama, writing a declaration calling for Ijaw control of the region's oil resources and environmental destiny. Within days the Nigerian military descended. Troops raided Oloibiri, where they shot the fifteen-year-old son of one of the kings while he slept.

Since 1999, the civilian government of President Olusegun Obasanjo has continued the pattern of violence. International human rights groups have documented nearly a dozen delta communities that have been destroyed with a mixture of killing, firebombing, and rape.

Just two months before my arrival, in February 2005, the delta village of Odioma, where a youth group was suspected of murdering a peace delegation from a nearby village, was firebombed and looted by the military, with as many as seventeen people killed. Reportedly, women were raped and the village elders were forced to eat sand. A few days after the destruction, the governor visited and pointedly said that the town should serve as a deterrent to other youth not to bring calamity to their communities.

The federal government's main concession to Ijaw complaints has been to give the five states of the delta more of the oil money, but that has worked to further disenfranchise the area. The delta states get 13 percent of the income from the oil they produce, many times the amount other states receive, and the money goes directly to the state governors, with no outside oversight. With so much money at stake, elections in the delta, which are slyly called "selections," are sewn up before they happen. There are no campaign speeches or carefully constructed constituencies. Instead politicians hire gangs of young men with guns to get the results they want.

"Nothing even faintly resembling an election took place in 2003. It was simply who had the most money and the most violence," says former human rights monitor Bronwyn Manby. "The continuing violence is tied up in the lack of elections, because everyone else is despaired. [The year] 2007 will be a bloodbath."

Violence here is useful, textured and nuanced. "The water project took Shell three to four years to complete," says one of Oloibiri's elders. "But if the youths do violence, then we get something." People refer to this formalized blackmail by the delicate term "youth restiveness." The elder stops for a moment, as if he's just realized how normal he's made this sound, then adds, "But we ask the youth to keep the peace."

While the elders give the tour, a small crowd of youths (which means any man under forty or so) follows at a discreet distance. One steps forward wearing a green shirt with a pattern of cocoa beans on it. He wants me to know that he'd prefer to support himself instead of living in his mother's house. His speech is cosmopolitan and precise. A World Bank study of delta families, which included Oloibiri, found they were spending 15 percent of their income on education—an enormous amount for people living on a dollar or two a day. But there's nothing in the village for this man. He'd like more opportunity, maybe a scholarship of the kind that Shell gives to youth in villages that are still producing oil. "An idle mind is the devil's workshop," he says, too seriously. I laugh—is that a threat? He repeats the phrase again, without any indication that it's funny. Maybe it's just a flat observation that his own mind has become preoccupied by devilish thoughts. Unemployment in delta villages is over 80 percent, and between 20,000 to 30,000 young men are in the militias or willing to join, making the devil's workshop one of the delta's largest employers.

The armature for the violence has arrived as different franchise opportunities to make money, the way one might move from selling Avon to selling real estate in the United States. Out of work, some young men formed groups called cults and fraternities with comic-book names: Icelanders, KKK, Bush Boys, Greenlanders, Black Axe, and "Burmuda Triangle, the unstoppable force." When politicians offered them money and weapons to enforce elections, some of them did.

And then came the oil bunkering. "If you make an impression, have stature and supremacy in an area, then a man will come and sell the initiative to you," explains a twenty-six-year-old political science student who is high up in one of the largest groups. The agent he saw, he says, was Lebanese, and he arrived with a package deal that included payoffs to the navy and weapons in exchange for the oil. Profits from bunkering, it was understood, accrued to the "political godfathers," the young man explained. He didn't want to name names; "for safety's sake, you don't say everything you know."

As the tour winds down, the elders lead me into a brick house with photos of a much younger J. C. Egba in formal clothes on the walls. They give me a soda and tell me what they'd like for the village. "We'd like it to be like your own place. A road, an embankment, micro-industries, transportation, and finance from aid agencies. Primary healthcare," one says. Shell, they say, should lead a "multipronged strategy" so that the government will fall in line, buttressed by international aid agencies.

An elder in a bowler hat says, "We are happy you're white. You will see and go directly so that people will come to help us. They should come directly." This must be the multipronged guilt strategy that reduced the ambassador to tears, but this seemingly rustic appeal has been planned with the same shrewdness that Washington insiders plot leaks to the media. International journalists in the delta automatically become part of a village-based PR strategy to embarrass the government or the oil companies into coughing up money. Information and violence are "power tools" here, explains a Western consultant. "There are layers of power and they're all shifting, so information is an important commodity—you can sell it; you can make people look stupid. The villagers are so fucking convincing. If they could keep themselves from being divided, they'd be very powerful."

Though the delta is remote and undeveloped, it's sometimes a shock-ingly postmodern place. More than four hundred years of trade with Eu-rope and the West—palm oil, slaves, and copra fibers—predated the oil trade. Everywhere there are reminders that the area was once industrial-ized and integrated with the global economy, used for resources, and then cut adrift. The delta's backwardness starts to feel like fast-forward when I literally stumble over a cannon hidden by foliage in another village. Be-hind it is a thick-walled holding pen from the slave trade, which started in the late 1400s here and didn't end until the early 1800s. Nearby a set of narrow-gauge train tracks used in the copra or palm oil trade leads to a wrecked jetty many times larger than anything here now.

While I'm inclined to see the oil business in postcolonial terms—as the actions of independent companies—I think the villagers see the oil companies as analogs of the Royal Niger Company, which had a Royal Charter to run an exploitive palm oil business and administer a brutal government in the delta between 1886 and 1900, with the blessing of the Queen of England. And perhaps the villages now see the American and European public as the beneficiaries and moral arbiters of the oil compa-nies and the Nigerian government and thus, like the queen, worthy of an appeal.

"Our daughter," says the elder in lavender, his eyes locking on mine. "Hundreds of people have come and nothing has happened."

**On the way home,** we pass one of the flares that dot the delta: A tongue of fire the size of a truck burns low and horizontal to the ground. Most places in the world have vertical flares, but Nigeria's flares present themselves in this particularly hellish way. Every day 2 billion cubic feet of natural gas go up in flames—energy that could be captured and used locally or sold. In 1996 the government asked Shell to shut down all of the flares by 2008, describing them as a fundamental violation of human rights. The company has said it will not meet that deadline. Reportedly, the flares are so big they can be seen from space, but they are still not big enough to get noticed.

**In Port Harcourt,** photos of militia leader Asari are on the covers of the local tabloids. "Asari going back to the creeks!" "Asari angry with government." In the yard where the buses congregate, men are selling eight-month-old magazines with Asari's belligerent, debauched face on them. Since the president signed a peace agreement with Asari and his rival the city has been fairly peaceful. The government has paid off all of the militias in the "creeks," which is how they describe the villages far from the city, and bought up AK-47s at about twice their street price. Despite daily, and credible, rumors that the groups are rearming, the government insists that all is well. One official tells the *Wall Street Journal* that the bunkering trade has been halved and he's locked up two hundred smugglers. In the delta, no one believes it.

Port Harcourt is a half-submerged wreck of a city where one traffic snarl leads to the next. Bits of expensive infrastructure—a pedestrian walkover, a six-lane highway, a corner of a park—poke out above a swarm of mold, foliage, algae, and more than a million people who've moved here to take advantage of the oil money.

Last fall Asari's men almost captured the governor's office. They would come into the city from the creeks in dozens of speedboats, take the ports, advance to within a block or so of the governor's office and then vanish back to the creeks. People in the port were unnerved. "The shelf life of the average Nigerian is falling," says a man who found bullets whizzing over his head while out for a walk with his wife. "Do you know why? Stress!" I want to see what the government makes of Asari's attacks, so I call the governor's chief of staff, and he agrees to talk with me immediately.

My only hope of getting to the interview through the midday traffic is to take a motorcycle taxi. No sooner have I hopped on the back than the driver pulls into the wrong lane against traffic, swerves around opening car doors, and hops a median. He cuts through a busy market lot, busting through groups of pedestrians with a honk. We slip around an open-pit sewer thick with mats of bubbling algae. Slithering between two moving cars, we meet another motorcycle taxi head on, play a game of chicken, win, and go on our way. The ride is a metaphor for life here, where simply surviving requires subverting the fallen infrastructure, grinding it even further into the ground.

The chief of staff's office features an oversize painting of himself wearing a white suit and white bowler with the Nigerian colors on a ribbon around the brim. Opposite this painting sits the man himself, a far smaller person with a round and comically amphibian face. Six cell phones and four buzzers are laid out in front of him on his large empty desk. The chief of staff is effervescent, regaling me with stories about his efforts to trace the drug trade in the delta, but he insists that Asari's militia didn't get anywhere near the government last fall. In fact, he says, there wasn't even really any fighting in the creeks. In his telling, it was all a publicity stunt. "Some NGOs were promoting the idea of the fighting," he says, "along with two newspapers. The NGOs put it on the Web to get money from donors, and there was an Ibo group in alliance with Asari and the Ogonis." This story is at odds with hundreds of newspaper accounts. Even his boss, the governor, claims to have purchased and destroyed more than 1,100 weapons as part of the peace agreement.

Talking with the chief of staff is maddening. One minute he is the oversize man in the painting, and the next he seems to shrivel in his chair, as if he's packed himself into a dot the size of an asterisk and left the scene. Obviously, the man I'm seeing—who has adopted a debonair tone and invited me to have brandy at his house that night—is just one manifestation of a far more complicated personage. Credible rumors connect him to support of political violence in the delta. Press accounts say that his personal driver was caught carrying a load of weapons to an arms dealer in another state, though the driver was killed while in jail overnight. The chief is both surreal and all too real.[1]

During my stay in Port Harcourt, I ran into one of the conflict consultants who shuttle between Chechnya and the delta. The consultant offers many dry and astonishing statements, cutting them with occasional anger. Political godfathers, he says, are playing a complicated and potentially catastrophic game with violence funded by oil bunkering. He offers these numbers: The cost of hiring a thousand trained militia members for a month and arming them with a thousand AK-47s is equivalent to the proceeds from just half a day's worth of bunkered oil in the delta. Before the 1999 and 2003 elections, bunkering spiked upward, providing money for weapons for the elections.

---

[1] The governor of the nearby state of Bayelsa was caught in London with a suitcase full of cash. He was later accused of helping himself to millions of dollars and buying an Ecuadorian oil refinery. He skipped bail in Great Britain and reportedly returned to Nigeria disguised as a woman.

I asked if the politicians were worried that they'd eventually lose the entire machine of oil and violence to war. "They are balancing the risk of arming the kids versus the risk of not being elected." It's a sentiment I hear all across the delta, where people repeatedly say that "It's risky not to take a risk."

Civil wars, the consultant says, are like bubble baths. "First there's water. Then you add soap and then the bubbles start—little ones at first. They get bigger and bigger and finally the bubbles join up and you have war. The bubbles here haven't joined up."

The self-destructive tendencies of Nigeria's government are hard to understand, and I find that categorizing it as a failed state doesn't help me put individual behavior in a larger context. But one evening I meet up with Nigerian author Ike Okonta. "A lot of ants but no anthill," he explains, waving his hands as the ants carry off the loot in different directions without the centripetal force, the organization, or the collective purpose offered by the anthill. He continues waving. "Just a lot of ants. Without a shared normative reality."

**Compared to the** chaos of Port Harcourt, the Shell compound is a great anthill of industrious behavior, furnished with carpets of shiny green grass. Once through several layers of security, I enter the clubhouse, where I meet my appointment, a senior Nigerian member of Shell's staff who works on implementing development projects. He welcomes me to sit on the patio by the pool to have a Coke.

To get to the pool, though, we have to walk through the Oloibiri Café, which is frostily air-conditioned and furnished with refrigerated glass cases containing fancy European chocolate cakes. The attendants are dressed in pastel uniforms, like nurses. My host seems surprised when I observe that the café is quite different from the village of Oloibiri.

"When they see Shell, they see the company as a debtor, owing them," he says. "We decided to let NGOs engage the community. When they see an independent body, they're quite appreciative, but when they see Shell . . ." He seems to feel that it is both quite unfair to Shell and completely normal.

He started his career as an engineer, joined Shell in the mid-1980s, and with the Ogoni crisis and the hanging of Ken Saro-Wiwa became involved in trying to reconsider the company's relationship to the communities of the delta. "We never claimed we were clean. The business of oil is not one where you can beat your chest and say it's clean. It's not. We're human and the pipelines have a range of life."

After Saro-Wiwa's death, international journalists pointed out the large number of oil spills in the delta (about six per week) and the sooty, archaic flares that burn off most of the natural gas associated with Nigeria's oil. They noted that soldiers had shot villagers from Shell's company helicopters. A boycott of Shell began in Europe and the United States, "The company stepped back and said why do the communities see us this way?" says the engineer. "The government is AWOL. The communities say we agree and accept this behavior because it allows us to do business."

Any discussion of the relationship between Shell and the Nigerian government inevitably becomes circular and convoluted. Shell's "Human Rights Dilemmas" workbook runs through the issue in different ways, cautioning "Shell is not the government," a phrase that has to be repeated so often because its meaning is so unclear here. Shell is the most robust and identifiable institution in Nigeria. Part of the reason human rights groups blame Shell when villagers are massacred is that Shell at least offers an entity to hold accountable.

But the complex relationship between Shell and the government has much larger dimensions. In 2004 Shell admitted that it had overestimated the size of its Nigerian oil reserves by 60 percent, not only to make its own reserves look better, but to help the Nigerian government increase the size of its OPEC quota.

The engineer remembers when he realized that Shell was being used as a messenger between the villages and the government. "The first time the [villagers] shut down the flow station in Nembe, we flew in in a helicopter and they were waiting at the flow station. We went to talk to the youths and women and they said, 'Go see the chiefs.' The chiefs said, 'Shell, we have no problem with you, but we need the government and we can't get the attention of the government. If you can't get the father you get the son.' "

Shell spent money on the communities as a calculated strategy to avoid losing oil production. But as it spent more, the violent attacks on the company rose. In the mid-1990s Shell said it spent $22 million on development and had about 100 sabotage attacks per year. Spending and attacks rose through the decade. "In 2000 we were reducing direct pay-ments and boosting development, but by the next year it was reversed and cash payouts were rising. Communities began to insist on cash. The cost of the community interface management had a steep rise—millions of dollars. It was staggering! And it was endangering the development program," the engineer says.

In 2003, when the situation appeared entirely unsustainable, Shell changed the name of the program from Community Development to Sustainable Community Development. In 2004 the company was spend-ing only $25 million on direct projects and giving another $68 million to a government development agency called the Niger Delta Development Commission. By then Shell had to break down violence into two cate-gories in its reports: criminal, which accounted for 314 incidents and 21 deaths, and community, which had 176 incidents.

One of the engineer's colleagues stops by the table to explain how the company felt it had no choice but to reward violence with cash. "It became a vicious circle. When people saw it was lucrative to go to the oil companies, they started having youth factions. Everything within the community got factionalized. Youth against chiefs." When the youth took over the flow stations, Shell ran the numbers. If the cost of the pay-off was 5 percent of production, then funds were made available. Angry villagers could cost the company millions by shutting down a flow station for a week. As the communities and the company struggled for control of the oil, development fell out of the picture. "The amount of money they spend on policing is enormous," he says. "Much more than the develop-ment projects." Recently both Shell and Chevron have publicly ac-knowledged their roles in nurturing conflict and dependency in the delta.

Since the beginning of the oil industry in Venezuela, multinational oil companies have been the screen between oil producers and oil con-sumers. When things go wrong, both sides blame the oil company. Sus-tainable development is an attractive idea to oil consumers, who'd prefer to think that oil production is benign, and it may remove some of the negative effects of making direct payouts to village chiefs, but on the ground it's less and less workable because it doesn't address the underly-ing issues: the broken state, the lack of jobs. In the delta, at least, Shell is a multinational at wits' end. The engineer prepares to head off across the big green lawn, saying "We're in a situation where people no longer see hope. And that is very dangerous."

Inside this compound, Shell has created a parallel Niger Delta that is prosperous, orderly, and gracious. This is the delta that would have been. Shell employees—who largely come from ethnic groups outside the delta—believe that they pulled themselves up by their bootstraps and hard work in school. Like Venezuela's oil workers, they are horrified and disgusted by the culture of dependency outside the company's gates. Here the streets are named after the different oil fields, and Oloibiri Street is lined with lovely little bungalows, their yards planted with flowers, their porches screened. In the mornings, employees go for an organized run with an ambulance following a few paces behind.

**Dennis sets** me up in an empty house that his employer plans to use as an office. The house is furnished with little more than a mattress and a succession of intimidating, welded steel, battering ram–proof doors. The electricity is not hooked up, so for the first few days Dennis, a former policeman, tries to get it connected. He is patient and bullheaded, and from some angles his bald pate and glasses make him look like a very exasperated Gandhi. He takes me to the national utility office, where we meet a man with pointy shoes and goatee who seems to accept payments in the waiting room. Back at the house, an electrician with a rickety ladder arrives, but by sundown there is still no electricity.

"If you push people beyond the threshold, you make them slaves of the moment. Tomorrow is a luxury," a Shell employee observed. "They are not networked into a larger system of hope. That works in the government's favor because there is a level of noncheck on public power. This place can't be Georgia or Kyrgyzstan [with their democratic revolutions]. By the end of the day you're knackered, just trying to cope."

It wasn't always like this. Down the street a former actress named Hilda Dokubo is starting a youth performance program called Street to Stars. (In the aftermath of the crisis with Asari, the government has funded several NGOs to work on peace issues in the delta.) Hilda is famous for her crying scenes on TV dramas. She is small, poised, and impatiently ferocious and speaks with a clipped British accent. It's quite hard to imagine her crying. In the 1970s, she says, there were schools and libraries and bus systems. The poverty rate was around 25 percent. "When I was in school we had clubs, but those are gone. The state library? You'd cry," she says.

She, and virtually everyone else, blames the end of Nigeria's middle class on the International Monetary Fund's Structural Adjustment Program in the 1980s, which required the government to cut services to qualify for debt assistance. The program is so notorious here that it's routinely referred to as SAP, and as in Venezuela, it is seen as a coup by IMF. "The SAP was when these institutions left, and under [military dictator] Ibrahim Babangida people fell from the middle class," she says. By 1985 poverty had doubled—to half the population. By 1996 two-thirds of the country lived in poverty. In 1997 some estimates were that 91 percent of the population lived on less than $2 a day. Encouraged to shrink, the government shriveled up altogether. A succession of military dictatorships survived by destroying whatever institutions remained while stashing billions of dollars in foreign banks. What was left was a void—a place where the anthill of the state used to be. Hilda dismisses sentiment. "Life has to work, if not peacefully, we must get it with violence."

There's a drive-by murder in front of the house where I'm staying. A former militia member was sprayed with bullets, they say, and a bystander got shot in the hand. Dennis goes nuts looking for me, worried that the police have taken me in as a "witness" in the hopes of extorting money. But at the time of the shooting, I was at a nearby Internet café, watching out of the corner of my eye as the two young men next to me crafted a classic Nigerian scam letter, complete with misspellings. In it they offered to sell the recipient a lost tanker of oil at a great discount if only the recipient sent them his bank account number and a wire transfer of $23,000.

If the bubbles in the bubble bath of war haven't joined up yet, it's because of people's fervent determination that they will not. As much as the tensions of oil are destabilizing the delta, there are still big reserves of hope here. A taxi driver tells me the problem with Asari is "a lack of love" on the part of the government. Then he gives me a flyer for a church sermon: "Impartation for Business Explosion." On Sunday I go to the service, slotted in with about two thousand people in a stifling white tent. The women are beautifully dressed, in prim cotton dresses elaborately decorated in sequins, beads, and piping. The preacher enters in a debonair white suit, and soon the crowd is screaming "Lord, I declare today my destiny will change." Business is about ideas and an exchange of values, the preacher says. "The reason Nigeria is a poor nation is because it doesn't exchange anything except oil." Ideas, he says, are "inspired dust." He quotes Tommy Hilfiger and Bill Gates, shames corrupt Nigerian presidents, rails against the wicked and the lazy. People seem eager for this dose of angry, hectoring hope with some Dale Carnegie thrown in.

Out in the creeks, Dennis takes me to ruined villages run by fierce young men who are waiting helplessly for government handouts. In Tombia, once a beautiful town of stone houses with carved wooden doors, a one-handed man named Prince Godo presides from a bombed-out building near the jetty. Godo is tall, with a fade haircut, and the fresh bandage on his wrist is a shocking white, even though he lost his hand during the fighting between Asari and the government months ago. He complains that the government isn't giving him the "empowerment" they promised during the amnesty. The village, which once held 44,000 people, is empty and being overgrown. A few old people creep among the ruined walls.

That night I meet a man originally from the village of Tombia who has a Ph.D. He is so exasperated he seems to have entered a permanently agitated state. "Where does this empty thinking come from? In Tombia everyone is blaming someone else. It is all gone!"

I ask him about Asari. "An errand boy. A pawn on a chessboard," he says derisively. "I knew him when his name was Dokubo Melford Goodhead Junior. He is not what we're looking for. Resource control. Shouting. Driving jeeps. Excuse me."

**Asari lives** in a nice part of Port Harcourt, where the dirt roads have been graded and channels have been dug to hold the rains. The houses here are surrounded by high walls, and diesel generators burp smoke out onto the street. Above his gate, two armed uniformed guards sit in a turret. They give me a piece of paper to write my business on, and after a few minutes they show me to a white plastic chair in the yard.

The days when Asari had three thousand men with him and thousands more ready to join are over. Five men wash his cars this morning—

the most skeletal of skeleton crews. People say that Asari's new wealth—profits from selling his weapons to the government, and payments and possible security contracts from the government and oil companies—sets him apart from his men. A month before I visited, Asari was reportedly receiving upwards of $7,000 per month from Shell for providing "security" and other services to the company. Shell said that Asari's company was not on its list of contractors.

There's a tendency to want to pin Asari down, to decide once and for all whether he's a gangster or a revolutionary. Bad or good? Part of the problem or part of the solution? It's even been suggested that he's on an Islamic jihad. But no simple description does him justice. "I'm not saying he's serious or frivolous," one person warned. "I'm saying he's complex." Master of the bubble bath, Asari is an exploiter of the area's violent infrastructure and enormous wealth for his own benefit. And this is not entirely a bad thing.

Nearly an hour passes until the big front door opens and a man stands in the doorway. He is young and pear-shaped, wearing a long buckwheat-color caftan. His face is smooth and pearish, too, and on another man it might appear soft. But he doesn't smile in greeting.

He shows me into his small office and sits heavily behind his desk. He checks his e-mail on his laptop, arranges his several cell phones, moves his stapler and tape dispenser. He's been described as a warlord, but I am meeting a successful entrepreneur.

He leans back in his chair and nearly closes his eyes; his apparent peacefulness comes across as calculating, as if he's secretly watching me through the slits. He says he is not involved in bunkering and implies that those who've said he is are lying.

"Ninety percent of our people believe in armed struggle," he says. "On November 23, 2003, I had just nine people. For nine months we were struggling, and we got volunteers every day. By the end of the struggle, we had 267,620 volunteers, including lawyers, boat mechanics, and doctors. They are disappointed that we haven't continued."

Asari says this in an angry but cool way, as if observing it from afar. Though he has called for bloodshed and been a party to it, there's something surreal about his presentation. Even in photographs from his nine months in the creeks, he's usually wearing his weapons uncomfortably, looking like a chubby kid who's accidentally ended up at the Special Forces summer camp instead of the one for video gamers. And maybe it's this separation that allows him to support violence as the path to change. "What I believe is that one gunshot is wordier than a thousand words."

Raised middle class, Asari attended law school but dropped out. "The key to him is anger. He was jobless and he began to drift. His anger then was mild. He had hope," says someone who knew him then. He was a searcher. He converted to Islam and took his family to Mecca, where they were so poor they paid for their trip by cooking Nigerian food on the street and selling it to other pilgrims.

Asari has spent more of his time as a creature of the system than an opponent of it. A local governor made him the leader of an Ijaw political group, giving him money and weapons and security contracts to enforce the 2003 election. Asari did all of this, then turned on his masters (it is said) when they didn't pay him off after the election. "After the election," he says, "I issued a statement that Obasanjo should return the stolen votes to the Ijaw people. This enraged the government." Whatever his motives, Asari's analysis was not far off. The delta delivered more than 90 percent of the vote to Obasanjo, and the region's rigged elections were responsible for a third of his winning margin of 12 million national votes, according to analysis by journalist Michael Peel.

After he denounced the leaders, Asari found himself on the run. The governor started supporting a powerful rival militia leader who attacked Asari everywhere, sometimes aided by the military. Asari began to describe his quest as an ethnic insurrection. He named his militia after Isaac Boro's original 1966 revolutionaries, and as we speak, he is trying to get a permit to hold a celebration in honor of his hero. Echoing Boro, he says that the delta was illegally conscripted by Nigeria at the time of independence.

I wondered if he had intended to spike the price of oil when he declared Operation Locust Feast in 2004. Was he emulating the villagers who kidnapped Shell's workers to get the government's attention on a grand scale? He says he was unaware of the price spike, and he certainly never intended it because an increase in oil prices would only benefit the government. "What we want is not publicity. We want results," he says. In fact, however, he's been far more adept at publicity—courting reporters and photographers—than at getting tangible results.

Asari applies violence bluntly, to get what he wants, but his goals are not long range. "He is not at all strategic," says Felix Tuodolo, a founder of the Ijaw Youth Congress. "If he were, things would be very different."

Asari seems to realize his value is as a wild card, not a revolutionary. "The government is afraid because I am a great mobilizer, a very emotive speaker, and whenever I talk people will search for something to do."

His most powerful argument is the unfairness of it all. "Is it another person's right to take the oil? The government is dictatorial and criminal to use the collective power of the people to dispossess citizens of this income. It's evil." I ask Asari about his vision for the future, and he says, "An Ijaw man with an Ijaw passport, not a slave passport. Ijaw currency, Ijaw language. I dream about that. I sleep at twelve o'clock every night, and if I can't sleep and my wives aren't here, I sit at the computer and I browse and I browse and I browse." It is a simple and plaintive vision— the insomniac militia leader surfing the Web, looking for ideas.

And that, oddly, makes Asari a hopeful sign. The movement for succession is not well developed, and it lacks big ideas and motivational speakers. Violence is mainly a paycheck for youth, not yet a cause itself. So far the violence directed at the government is mainly to facilitate dialogue—perverse as that may seem—and not to start a war. And that is why Asari was able to bring the area to the edge of war—just before the bubbles joined up—and then call it off.

The desires of the people living next to the wells are simple still: water, electricity, jobs, education. If Hugo Chávez were here, he'd send in some Cuban doctors and some cheap frozen chickens and stir up some anti-American sentiment. But the Nigerian government can't seem to muster anything remotely close to that.

And so in this period of lull, Asari plays a deceptively positive role, threatening the government so that it is roused just enough to deliver on its promises to the delta. "He takes a carrot-and-stick approach," says Patterson Ogun, who disagrees with Asari's commitment to violence. "He's operating on several different levels. The lesson is that the federal government and the oil companies are deaf, dumb, and blind until there are glaring issues of insecurity in their faces. You have to apply some minimum amount of force to get an ear." More bluntly, another Ijaw activist who is ambivalent about Asari's methods describes him as "insurance against genocide."


**I am knackered.** The electricity is on and then off. When off, the generator provides a few hours of light at night. Then the generator breaks. I stockpile water around the house in rows of plastic bottles. The power is reconnected, but the current is so faint the fan barely turns. Mysteriously, the plumbing seems to be electrified. One morning when I'm washing my clothes in the sink, both arms touch the faucets and I get a shock. And then, after nearly three weeks, the junction box on the house smokes, flames thinly, and goes dead. So that's the end of the electricity.

The U.S. strategy for oil has been to count on the distances between producers and consumers to keep the trouble and the burdens of oil production from being our business. But that system has brought about its own demise. Oil has made the world a smaller place, and now globalization magnifies events at the wellheads, transmitting them instantly to the NYMEX. When Asari threatened the government in the fall of 2004, his actions removed a mere 30,000 barrels a day from the market (irrelevant even in Nigeria's 2.5 million barrels in exports), but the effect at New York's NYMEX was huge. Averaged across all of the barrels traded that day, Asari's threat may have cost the world an extra $100 million. In a tight oil market, the globe is a set of funhouse mirrors, and the price of your next tank of gas is only a cell phone call away.

**I have one** more village to visit. Dennis takes me on an hours-long boat ride into the creeks to a community that sits on a plot of land about the size of three basketball courts hollowed out of the mangrove swamp. The military destroyed the village when Asari was on the run in 2004, and now about two hundred youths live there in small wooden stalls with dirt floors and flimsy dividers between one spot and the next. One of a handful of women in the community fans herself slowly while drinking a beer. She's wearing an extravagant dress made of pistachio-color lace, with floaty sleeves. Around her, the ground glints with whole and broken gin bottles. The town flag has a machine gun on it. When I try to take a picture, the spokesman, a young man named Mansen, stops me, saying people will get the wrong impression that the town is full of "hoodlums."

Mansen is carrying a book called *Better Life from Oil Wealth: A Discussion Guide for Public Forums on Conflict Associated with Natural Resources.* He manages to work the phrase "the inexplicable loss of biodiversity in the delta since the onset of oil production" into the conversation more than once.

Mansen claims the men are all fishermen, though the docking area is filled with spilled oil and there is not a fishing net (or a fish) in sight. He also claims that the oil industry has destroyed all the fish. My impression is that the village survives by bunkering or providing security for bunkerers. The creeks are crawling with low-slung boats jammed with what look like huge plastic urns—each as tall as a man—all filled with crude or other oil products removed from the pipelines.

But for a village with a lot of money, the mood is despondent. The men carry expensive cell phones clipped to their belts even though there is no signal for an hour's ride in any direction. They seem to want to be part of the world but feel hopelessly cut off and drifting farther away. Or, as Mansen puts it, "It is very highly regrettable and condemnable for an oil-bearing community to be so marginalized."

More even than Oloibiri, this village is connected to the world only by the slender stream of oil that it controls, and it is driven mad by the implications. These restless men with their AK-47s, useless cell phones, and lists of environmental talking points defy categorization. U.S. think-tankers are fretting over whether they're "legitimate opposition or criminals." But we really don't have the right words for this. Rebelusionaries? Envirowarlords? They describe themselves as victims, but that may change.

Dennis is depressed by the village. Usually a stoic, he clucks his tongue and shakes his head. "Did you see their women? As beautiful as movie stars," he says before trying to find some biscuits. Ten minutes later, he comes back. No biscuits, only beer, soda, and gin.

Mansen rides the boat back to town with us. "When you are so marginalized and when dialogue fails," he says, "there is another option. And that is what we find ourselves in today. Dialogue in the Nigerian context has been counterproductive. Instead of sitting and dying, we'll go out and die [violently] and find solace." I gather he intends this as a threat against the Nigerian state, but it sounds as if he's aiming to prevent genocide by committing mass suicide.

Back in Port Harcourt, I meet up with Dimieari Von Kemedi, an Ijaw activist who helped negotiate the peace deal between Asari and the government in the fall of 2004. Von gives me a ride around Port Harcourt in his car and I ask him when he thinks the violence will stop. "When this hydrogen thing takes off and all the oil isn't worth anything," he quips.

Then he gets serious. "This whole thing about suffering and marginalization is random disorganized thinking. People are talking about empowerment and resource control without knowing the meaning. These terms are sandbags." The problem isn't the oil itself, but the relationships that have been created around it. The entire project needs to be rethought, each relationship reconsidered, by people inside and outside the delta.

"What people don't realize is that everything starts with ideas. Everything! Democracy is an idea!" He taps his steering wheel. "This car started as an idea! If you want to get anything done you need ideas. And what it will take in the delta is men and women with ideas debating, strategizing, and using tactics systematically, working through the different angles." He continues driving, and as we approach a roundabout we barely escape being smashed by an out-of-control truck. I turn around to look at it careening into the intersection, but Von continues unfazed. "You have to be properly depressed to talk about solutions," he says.

**The oil companies** are starting to get depressed, but they know that the cost of fixing the delta—even if for no more humanitarian purpose than ensuring the outflow of oil—is enormous. And that's assuming it can be done at all. Shell has hired a strategist, formerly with the UN, who lives in the compound on Oloibiri Street. He acknowledges that the company's development projects actually contribute to dependency and violence in the delta, scoffing at the company's meticulous audit that found that between 50 and 86 percent of recent development projects are "successful." He names a village I've been to, one where Shell constructed an impressive concrete town center and people live in shacks on the edges. "People say that is a model," he says, "but we're deceiving ourselves. If you go there in the morning, the youth are doing precisely nothing! They should be producing. Successful development is people being active economic agents."

In his opinion, part of the problem is that international governments are more comfortable using oil companies as foils than getting involved in the sticky business of making real change in the delta. "International governments can bash Shell all they want, but someone needs to kick the ass of the Nigerian government! The absence of government is 80 percent of the problem. The donors like the World Bank, the IMF, the

USAID should get over the bullshit and start intervening on behalf of poor people. They should come in and do development here with the same ruthlessness and efficiency that they enforced the Structural Adjustment Programs of the 1980s. If I had my way, I'd wind down Shell Community Development. Start next year with something called the Niger Delta Trust with the oil companies as trustees and the aid agencies as partners and cover the whole Niger Delta. This should have started happening yesterday." He finishes the rant and takes a drink of water, blinking. Through his screened porch, the flowers behind him look like an oil painting, an impossibly pastoral dream.

Other delta activists believe that a major development program is not in the cards. "If there's a political crisis, you pack your bags and go home," says one. "Nobody here wants to spend money on community development when you could militarize the area." To some observers, that appears to be what's happening. The United States has given the Nigerian navy several coast guard cutters. The FBI has offered Nigerian corruption fighters training in preventing oil theft under antiterrorism funding. The area has been described as "of strategic interest" to the United States, which implies that the military would step in if oil were endangered. In early 2006 the vice president of Nigeria, Abubakar Atiku, told a reporter for Britain's *Financial Times* that he wished the U.S. military would intervene to stop fighting in the Niger Delta. Later he denied saying this.

Surprisingly, support for a big development program comes from a senior official at the U.S. Defense Department's center for strategic studies in Africa. "There is no military solution to the problems in the delta," he says. "Bunkering changes the way we need to think about problems. It's a change from the oil diplomacy of twenty years ago. Before it was unidimensional—instability requires intervention. Now we need to look at more than one instrument of U.S. national power. We cannot address security without addressing development. We need a broader definition of security—welfare of the individual, physical security, and environmental protection. We need to coordinate peacekeeping, development, and capacity building. That hasn't happened in the past. We need to coordinate with the EU [European Union] and China as well as the oil companies." I'm surprised to be hearing this from a security strategist, whom I'd expect to talk about short-term military engagement rather than the most expensive development program in the world. Is this proposal taken seriously? "It's more of a wish at the moment," he says with a laugh.

Offering more than money and promises may be inevitable, from a competitive standpoint, because China is now offering development projects as part of its investment plans. In 2005 China invested $3 billion in Nigerian oil fields and promised to build a refinery to make fuel for the Nigerian market.

And the calculus of the cost of violence in Nigeria is shifting with every new price spike. "Last year people said debt relief was too expensive," says a senior U.S. Africa policy advisor. "But what's changed is the price of oil has gone up and the U.S. has no excess capacity. The biggest change is the U.S. military who is starting to say that they're going to have to go in and clean this up. This administration says it'll take a million years to fix Nigeria. I think it could take twenty or thirty. It is possible to change the culture."

In the delta kingdom of Akassa, the Norwegian oil company Statoil has been funding a development project for nineteen communities totaling 180,000 people since 1997, administered by an NGO called Pro-Natura International. The primary task of the project has been to build local institutions: microcredit associations, women's groups, small businesses, and associations to build infrastructure, preserve the environment, and manage community funds. Akassa is dotted with the usual mixture of abandoned development projects—the sagging schools and empty hospitals left by other oil companies—but here several of them are in use, restored with the locals' own labor and at their discretion. Akassa still has a lot of problems—many of the youth are still unemployed, and as other oil companies enter the kingdom, communities are starting to turn on each other—but compared to elsewhere in the delta, people have some faith in their local institutions, and they are willing to talk about the future without talking about oil. "I'm expecting the community will want to build a road," says one of the chiefs. "And then you can drive to another place in the twinkle of an eye. That will change people's lives—there will be jobs, even people selling tires for the cars, and Akassa will become a main town."

**I go north** to see if Nigeria's federal government has a similar vision. In the capital of Abuja, oil money has built a grand fantasy of Nigerian federalism, with a pastiche of architectural styles from the three main ethnic groups. The Secretariat, for example, is a swelling pink pudding topped with green domes. One legislator's home—a miniature version of Tara—has a masonry model of an airplane—all fat and cartoonish—plastered on top. Here the very idea of accountability is undercut by the street names, which still honor brutal dictators like Sani Abacha. I travel through ever more elaborate gates, and past roaming peacocks, to reach the office of presidential advisor Oby Ezekwesili.

Ezekwesili is tall and formidable, with an aura of charisma and power. She wears a tailored batik suit and spike heels and enters her office with her large shoulders slumped, lugging two bulging bags of papers. Oil has brought "delusions of grandeur" and tragic expectations for rich and poor alike, she says.

She has a rich, throaty voice and an off-putting delivery. "I had Shell executives sitting here in my office and I said the problem with the oil sector in Nigeria is you guys are bloody arrogant! You need to get broken, make yourself vulnerable! Show the communities you feel pain." Her story crescendos before I'm able to ask how she expects a multinational to feel pain. "Maybe they call it soft power," she continues. "Maybe you should leave off all the strength you have and maybe you should go"—she holds here for a dramatic moment—"just watch a mooovie." This takes me off guard, and she laughs uproariously at my reaction. Shell, she says, is shirking its responsibilities, as are the state governors.

What I don't get from Ezekwesili is a vision, even a simple one, of how the government intends to engage with the people of the delta to build a future together. Maybe that is because the international community—and the oil consumers—don't ask that of the Nigerian government. In late 2005, Nigeria was able to cut a deal to pay off $12 billion in foreign debts (using its windfall from high oil prices) and have $17 billion forgiven, in exchange for promising no more than macroeconomic stability. The creditors missed the opportunity to ask for change in the delta; they just canceled the debt and appeared to endorse the status quo.

I ask her if she sees a possibility of losing the whole Niger Delta into its institutions of violence. "At times I look at the rascality that we still deal with. And I say what are you doing? It's only been six years. This is a hangover of military rule. With more elections let democracy grow on us." What if, I persist, you don't have that much time? "It's amazing how things work out in Nigeria," she says with a less than friendly glint, preparing to dismiss me. "I think the Niger Delta will work out."

**The bubbles** continue to form. In the fall of 2005 the federal government jailed Asari on treason charges, but he still remains a force in delta politics. In early 2006 a much more violent and better-organized militia group named Movement for the Emancipation of the Niger Delta began kidnapping foreign oil workers and attacking oil infrastructure. It demanded the release of Asari and seemed to be a more lethal version of what had gone before, with connections to powerfully placed politicians. Bigger bubbles, but still not joined up. While the kidnappers were issuing ultimatums, Nigerian president Obasanjo appeared at the World Economic Forum in Davos, Switzerland, pushing his country's tourism industry. By 2006, as violence worsened, he announced a massive delta-wide development project to provide roads, water, electricity, tens of thousands of jobs, and better cell phone reception.

## 11 ⸳ CHINA  *DRIVING THE "ASPIRE"*

Stuck in Beijing noontime traffic. A man walks among the cars selling turtles from a burlap bag. Another follows selling small, applelike hawthorns. When I roll down the window I can smell the minty-pungent horse standing next to us. It's pulling a cart with a red-cheeked woman in a pink jacket holding the reins. She looks exactly like a model peasant in the old propaganda posters. Life in China, where GDP has been growing by 9 percent a year for more than a decade, is one big *whoosh*, a constant struggle for bearings, even in a traffic jam. Idly, I wonder when the capital will get around to banning horsecarts.

On Beijing's skyline, the big tussle between authoritarianism and go-go growth is playing itself out. The squat, amphibian brick monoliths that signaled one-party architecture and one-party rule are being shown up by eccentric stalks of glass, metal, and granite. The city is changing so fast they're publishing new maps every week. Hood ornaments are important here, a designer says, because they remind the driver of his status, sort of a portable landmark in a constantly shifting landscape. Statistically, half of the people leaning on their horns didn't even drive four years ago.

Mao famously said that revolution occurs only at the end of a gun, but this one is happening at the wheel of a car as millions of Chinese start driving. The effects of this revolution in personal mobility are being felt in the world oil market, where Chinese oil consumption has risen 45 percent in the last five years, in geopolitics, and at home in China, where it's challenging the leadership of the Communist Party. But that's only the beginning of the revolution, because now China is also making the cars. For the past century cars have been a quintessentially American thing, produced for the big and growing U.S. market and exported to people who wanted to adopt the American lifestyle. But the U.S. market is now so saturated, the only way to sell more cars is to lower the driving age. The next big market is in China, which has the same number of cars per capita as the United States did in 1910—and that means the future is here, in China, where the car is starting its second life.

After months of traveling in oil-producing states with their stunted economies, I'm almost enjoying the smog, the furious activity, the feeling that even as I sit in traffic, China itself is hurtling somewhere. In a few days, when I have a nasty lung infection, my sentiments will have changed.

The smog and the traffic jam are a direct result of the booming economy. As China's GDP hit $1,000 per capita, people began to buy autos. In 2000, there were 16 million cars in China; by 2004 there were 27 million. By 2010, maybe 56 million, doubling by 2020 to 120 million, all crowding onto the roads. "When a car becomes something everyone can afford, forget it," says one driver. "You won't be able to drive."

And all the cars are burning oil. In 1992 China exported oil. But one year later it started importing, and each year since 1993 it's needed more. In 2003 China passed Japan to became the world's second largest oil importer. In 2004 it needed nearly an extra million barrels more oil per day than in 2003. (The culprit was the diesel generators people set up to deal with electrical blackouts and brownouts.) In 2005 the electrical situation was on the mend, but the country still needed an extra 230,000 barrels a day—the equivalent of another Chad. And so on, as oil feeds China's GDP and the GDP drives the desire for cars that use more oil. By 2025, perhaps, China will import as much crude as the United States does now.

For the first time, the United States has to consider a rival energy consumer as hungry as itself, though not as rich. In New York, oil traders have blamed record-high oil prices on China's demand.

China sees its need for oil as both a strategic challenge and a business opportunity. In 2005 China set about signing long-term supply contracts all over the world. In Iran, it promised $70 billion and an auto assembly plant. In Nigeria, $3 billion and a refinery. Saudi Arabia, Kazakhstan, Russia, Canada, Ecuador, Angola, Chad, Syria. One of the Chinese national oil companies tried to buy the American oil company Unocal but was blocked by an indignant Congress. China made it clear it was more interested in getting oil, in just doing business, than in getting involved in the politics of oil producers. But being "apolitical" is just another kind of politics. When the UN Security Council voted on whether to sanction Sudan for genocide in Darfur, China, which buys half of Sudan's oil, blocked the move. History was repeating itself. In the 1940s, the United States took over Britain's old oil concessions, claiming it would just "do business." Now it's China's turn.

"We have been producing and exporting oil for more than 100 years. But these have been 100 years of domination by the United States," said Hugo Chávez, after signing an oil deal with China that included a $700 million credit line for low-income housing. "Now we are free, and place this oil at the disposal of the great Chinese fatherland."

**Here in** the Great Fatherland, though, the car culture is an American import, even if the cars themselves aren't. Ellen has razor-short hair, a quirky style, and a dreamy edge. She works for a foreign company. "The car is the first level of Western life," she says, admitting that she learned to drive because she'd watched "too many American movies." Cars for her were about some idea of being outside herself. "In a car I imagined I'd be like an astronaut: free. I was hoping to just concentrate on what's ahead and forget about reality."

So she took driving lessons, which were fairly expensive and required that she pay an "extra fee" to the instructor so that she could actually get behind the wheel of a car. She didn't get an "American" feeling when she drove on China's roads, though. Once she was driving, she never managed to go faster than 60 mph. "It wasn't a feeling of freedom," she says. "The people beside me seemed worried."

I wanted to know more about the people in the traffic jam, so Ellen and her friend Jessie (they use English names at the joint-venture companies where they work) took me to an auto showroom in the suburbs, where a rigid grid of streets met some old fields.

The Hyundai showroom is a smallish glass room containing five cars and about twenty extended families. A tiny saleswoman in a cheetah-print skirt is acting as the ringmaster.

The families, mostly young couples with both sets of parents, are shopping for their first cars. They poke at the stitching on the seats, evaluate the design of the rear lights. People seem to appreciate cars that have a "face" made from the lights and the front grille. The young people spend a lot of time sitting in the rear seats imagining how their parents will feel. I've never seen anyone test a car this way.

No one is interested in talking about fuel economy or traffic. As one car owner explained to me later, "Oil is none of my business. That's the structure of society. For five thousand years the government has been the emperor and he should do everything. I don't expect to participate in the decision-making process."

We end up sitting in some chairs at the side of the showroom, people watching. Near us a TV is on. A woman in a yellow dress is twirling in front of a suburban house with a grassy yard on an empty street. It looks like an advertisement for allergy medicine, but apparently it's from the city government. I ask Jessie to translate the song. "Nowadays life is getting better and better, sweeter and sweeter. Everyone can fulfill their dreams. The roads are getting wider and wider. Let's work together to build a brighter tomorrow."

Of the two women, Jessie is more conservative, with longer hair and glasses. She has a determined set to her face as she explains why she expects to get a car. "I'm doing what I'm supposed to, going along with the economic model that the government has designed," she says with a bit of force. "They set up a system. I do what I can to achieve. If I become a manager, I expect to have a car. A car now is an expected thing." The car is part of her unspoken contract with her company, the government, everyone. She says it would allow her and her husband to take better care of their parents.

Ellen eyes the ad and seems annoyed by its promises. We continue to watch the couples shopping for cars. The little saleswoman in the cheetah skirt is loudly bawling out a worker. Ellen says, "I'm thinking about taking flying lessons."


**On the other** side of the Beijing traffic jam, I meet one of the men in charge of managing and encouraging economic growth. Dr. Yang Yiyong wears a working bureaucrat's pinstripe suit, glasses, and a tired look. His background is in economics, sociology, and anthropology, he mentions. And now he is the deputy director general of the Institute of Economic Research, at the National Development and Reform Commission, the think tank for the Communist Party's leadership. He's high enough in the hierarchy that he prefers to meet out of sight in a private basement dining room of a restaurant known for duck with stewed cherries, low enough that he'll speak to a foreign reporter at all.

He starts off by saying that he doesn't like cars. "I object to this vague notion of status," he says, laughing a bit at himself. More seriously, he says, people worry that promoting American ways of life could create a disaster in China. But it doesn't really matter how he feels about them: Cars are here to stay.

Yang starts by laying out the numbers that govern his job: China's projected population peak of 1.5 billion in 2030 requires continuing GDP growth of 8 or 9 percent per year. And because China's industry is inefficient, it requires massive amounts of energy. In 2005 China's oil imports rose by nearly 16 percent, faster than GDP. Oil imports alone will need to quadruple by 2020. "There is no hope of lessening energy requirements until 2100, when the population will drop below a billion." I've never heard an American planner talk in these time frames. It's a bit thrilling.

Oil has become the second most important issue for government economists, right after food. Picking at his duck, Yang explains the urgency. Without strict controls, China's growth will be choked by lack of energy, and because the country is not as efficient as, say, Japan, it requires four or five times as much energy for a single point of GDP. If the GDP doesn't grow, there will be unemployment and unrest. So energy is directly related to "stability," aka the government's hold on power to 2100.

Oil, in Yang's opinion, inevitably leads to war, and he is joined in this belief by every other official I speak with. Yang uses a pun to summarize the leadership's view: "If you pump oil, you have to fight for it." ("Pump" and "fight" sound similar in Chinese.) "War may not be a civilized solution," he says, "but it's effective." The intent of the U.S. war in Iraq, he believes, is to fund itself through the cheaper oil that may come later. "These days if there's no oil, there's no country and no future for the United States."

Yang says China needs "a bigger space to survive under U.S. hegemony." So China is focusing on buying oil but also on conserving it, becoming more efficient, and developing new fuels. "We're not saying we can reduce consumption, but we can reduce the increase to win time."

Other observers, particularly European analyst Jonathan Story, believe that U.S. hegemony is particularly useful for China and an implicit part of China's overall strategy. If China can rely on the United States to bear the expense of maintaining the energy status quo around the world for the next decade or two, it will be able to retool its economy, "and then the world's balance of power is forever altered." In other words, Story says, lack of change in the short term has revolutionary implications in the long term.

**In 2004** China shook automakers worldwide with the incredible speed and strictness of the auto fuel efficiency standards it enacted, which are 5 to 10 percent stricter than U.S. standards and among the toughest in the world.

The task of writing the rules fell to the Ministry of Standards, which sits in a newly redeveloped section of Beijing. "We're next to the two buildings that look like syringes," Yin Minhan, director of the Department of Industry and Transportation, says with a laugh. Sure enough, there are two looming buildings shaped like syringes, made from Kafkaesque green stone. Yin himself is the epitome of bright efficiency: Since 2000 he's worked on energy standards for a fast-forward social history of Chinese consumerism: first electric motors, then refrigerators, air conditioners, and now cars.

Yin's group met with consultants from China as well as the Energy Foundation, a U.S.-funded NGO, and then traveled to Japan, the United States, and Europe to gather opinions on efficiency regimes. They decided to create a scheme that rewarded smaller cars and imposed stricter fuel efficiency standards on larger ones. The standards go into effect in two stages. In the first stage, only one U.S.-made SUV passes. The second stage is harder still. "We learned our lessons from the U.S.," says Wang Junwei, one of the five hundred or so people involved in auto standards at the ministry. "We are going to clamp down on SUVs early!"

But there was a bigger strategy behind the rules than merely saving fuel and preventing pollution. The ultimate intent of the regulations was to make Chinese-built cars more exportable to high-end markets, such as Europe. Designed to pressure joint ventures like GM and Volkswagen to send their newest technology to China, the standards are part of the slow revolution that could make China the new Detroit.

When the Chinese bureaucrats in charge of the standards listened to

Detroit auto executives denigrate fuel economy standards, they heard an opportunity. The team perceived Detroit's reluctance as a strategic weakness and a clear way for China's industry to become more competitive. "China doesn't subscribe to the idea that what's good for GM is good for the country," an American consultant who worked with the government team says with a laugh.

**Another afternoon:** a mad scramble across Beijing to get to the pink granite office tower where Dr. Zhou Dadi works. The elevator to his office in the tower smells like duty-free perfume, but Zhou, who is the director of the National Development and Reform Commission's Energy Research Institute, adheres to a more spartan standard. His office is decorated with admonitions to turn off the lights. He tells me to put my coat on because the heat will not be on for another week. The only decoration is a glass polar bear that starts to resemble an icicle while we talk. "There are no examples of a rise in GDP with lower energy use," he says grimly. "We need energy efficiency."

Zhou started out working on greenhouse gas issues in the early 1990s, developing models of the impact of climate change in China, which will be significant. Making the country more efficient, he reasoned, would accomplish three goals: reduce energy use, reduce pollution and greenhouse gases, and make China's industry more competitive. In the late 1990s, Zhou worked on market reforms to close inefficient factories fueled by coal. Burning China's huge coal reserves for power makes the country's greenhouse gas emissions per unit of GDP sixty-eight times that of Japan and six times the rate of the United States.

"Although we have the political willingness to reduce energy use, it will not be easy," he says. "We still need to make the market mechanisms work in this time of globalization." Zhou thinks that the developed world should set a better example in cutting back on greenhouse emissions. China, he says, can't live like the West, with lots of cars and waste. "We have to eliminate all bad practices."

Zhou says that the goal is to replace 10 percent of China's energy use with renewable sources by 2020, with the final aim of reducing oil imports to just 6 percent of consumption. But by late 2005, after we had spoken, that target had shifted upward. Now the country was going to get

15 percent of its energy from renewables by 2020. And in early 2006 Premier Wen Jiabao announced that energy efficiency was a new economic marker, and he expected China's energy use per unit of GDP to start falling by 4 percent a year.

"We still don't know if these measures are enough to change the market," says Zhou, "but I don't think we can wait. This will take twenty to thirty years, and we will keep studying it."

**Shanghai, 8:07 A.M.** Taxi caught in a dense felt of vehicles moving herky jerky out of the city toward the suburbs, horns blaring. How fast are we going? It's impossible to tell. On days clear enough to see the horizon, Shanghai's new skyscrapers are a boisterous bunch, their tops ornamented with abstract symbols that look like kitchen implements: a cheese grater, salt shaker, egg timer, toast tongs, a juicer—all celebrations of the triumph of pragmatism. Shanghai's smog, caused mostly by cars, hangs low today, bringing the horizon to the edge of the road, where the bicycles are. Lanes? Nobody bothers. Next to me, a monstrous pale blue Liberation truck studded with 1930s rivets the size of soda caps breathes its exhaust on the latest sleek eggshell Audi. On the other side, a sinew of a man pedals a freight tricycle glopping gobs of used cooking fat.

For all the logic of Beijing's top-down decision making, there's no guarantee the goals will be followed at the local level. Even the efficiency standards, which seem clear enough, may be ignored. Governments of some cities, such as Guangzhou, have forbidden small cars from their streets on the grounds that they are beneath the dignity of a metropolis.

The usual methods of traffic control may not sway the nouveau riche and their habits of conspicuous consumption. A city planner frets that even high tolls will not cut the congestion. "The rich people won't care if you charge them to get into the city because they need to show people they have a car." In China, even tolls could too easily become status symbols.

And the things the city had going for it—all those bicycles, for instance—are rapidly disappearing. Shanghai's narrow streets have become so busy that commuters have started crowding onto the subway just to avoid the cars. Traffic deaths in China, caused in part by all of these new drivers, grew to the equivalent of a daily 747 crash.

In the edgy logic of China's traffic jam, it's not clear how this battle will sort itself out. "Everyone's mind is changing," says Dr. He Dong-chuan, of the U.S. NGO, the Energy Foundation in China. "There's controversy over how to grow, and about green GDP." What combination of strategy and accident, idealism and calamity will influence the future of the car here?

Forty minutes past the traffic jam sit six square miles named Shanghai International Auto City, recently carved from the rice fields of a town called Jiading. Three years ago Shanghai decided it wanted to build a place for its auto industry to become the largest in the world. Out went Jiading's farmers and little factories. In went Tongji University's College of Automotive Studies, spaces for joint-venture auto assembly plants, parts suppliers, testing facilities, a car museum, a wind tunnel, a golf course, and a $320 million state-of-the-art Formula One track—in the shape of the first character of Shanghai's name, which means, roughly, "upward."

This morning in October 2004, the Formula One track has been taken over by a swarm of hydrogen fuel-cell vehicles, along with electric cars, natural gas buses, and a very stinky little car that claims to run on hydrogen peroxide. The fuel cells come from the biggest of the first world's big—Daimler Chrysler, Toyota, GM, Ford, and the rest—which have come to compete in the Challenge Bibendum, a competition for alternative fuel and low emissions vehicles sponsored by Michelin. (Bibendum is the name of Michelin's morbidly obese tire-man mascot.)

The big draw is not the competition but the venue. China's government has signaled that it might be willing to make policies and spend money to make the car of the future happen here faster. In the rest of the world, that's still only a dream.

But in China, low-polluting cars may be the pragmatic option. For one thing, pollution is choking off as much as 15 percent of GDP growth by causing lung illnesses and other losses. That makes clean vehicles look almost practical. And because of this, Chinese cities and taxi companies are buying. "This will be the biggest market in the world by 2010," says Gilles Debonnet, director general of Citroën China, as he stands next to a CNG (compressed natural gas) taxi designed for Beijing. "If we don't bring a [low-emissions] solution to the taxi market then we can't stay." Toyota offered to produce its new hybrid Prius in China even though it wasn't clear whether the relatively expensive car would be popular.

And China's other strength is all of those bicycles. Gasoline-fueled cars are not fully entrenched here, as they are in the United States and Europe, and a radically new technology, such as hydrogen, might take off (with a lot of government help, of course). General Motors hopes so. "We're trying to engage the government," GM China VP David Chen says as he runs after a Shanghai government official at the track. "Our theorists believe China has an advantage with fuel cells because it has no infrastructure and no resistance. It's been cut off from the world for thirty years. It may be a unique situation to leapfrog."

Also lurking around the rally is Shell Hydrogen VP Gabriel de Sheemaker, who worked installing Iceland's huge hydrogen experiment and is now at Bibendum trying to scale up. "We consider China a wild card," he says. He's looked at the $1 billion that Shanghai spent on a German maglev train and seen possibilities. De Sheemaker's eyes get dreamy as he imagines Shanghai on hydrogen—city blocks powered by fuel cells, cars filled from hydrogen supplies embedded in the buildings. "In Deng's day he experimented with whole cities!"

Hydrogen is attractive because it produces no greenhouse gases or pollution when it's used and because it can be produced from a variety of sources, including oil, gas, nuclear, solar, biomass, and coal. Hydrogen is bedeviled by several major problems, though. The first is the so-called chicken-egg problem: Hydrogen requires both entirely new cars and a new fueling infrastructure, and both must arrive at the same time. Hydrogen cars generally run on fuel cells, which create electricity through a chemical reaction, but no fuel cells are commercially viable yet. A hydrogen fueling infrastructure would be very expensive, requiring thousands of new facilities equipped to make, hold, and pressurize hydrogen gas. The second problem is the fuel itself: Hydrogen is difficult to store and expensive to produce. At the moment, hydrogen vehicles actually consume more fossil fuels than regular cars, because the process of converting natural gas to hydrogen involves the loss of about a third of the energy contained in the natural gas.

As a result many of the cars at Bibendum are defensive about their impracticality. You'd think that a show of the environmentally friendly future cars would be like visiting a parking lot for spaceships. But because the cars are struggling to be taken seriously, they've gone to great lengths to disguise their billion-dollar fuel-cell guts under the frumpiest possible exteriors. Mercedes-Benz has stuffed the magnificent insides of its A-Class F-Cell into a blah hatchback. Ford has deposited its fuel cell into the unassuming body of a Focus station wagon. Hyundai's is in an SUV with a butterfly decal.

Amid this expensive ordinariness sits a bizarre little yellow car shaped like an egg. More specifically, it looks like a large shiny yellow tadpole that has swallowed a very tiny pickup truck. In front: seats for two behind a window shaped like ski goggles. Viewed head on, the headlights and motor cover make big cartoonish eyes and a grin. Car as smiley face. I feel silly grinning back at it. In back, a cute shelf, like a very abbreviated truck bed, with rocket-shape taillights. The people who made the car are students from the Wuhan Institute of Technology, and they're hanging out in a giddy gaggle near the car. One runs to get their professor. The name of the car is the Aspire, they say, because it's "the spirit."

The student designer, a sincere twenty-one-year-old in a white shirt, says the car is designed to attract "youngbloods" and "new white collars," which is how he characterizes the aspiring global middle class—that vast crowd who is just getting enough money to think about buying a car but whose hopes are constrained by traffic, pollution, and potential fuel shortages. Car owners currently make up about 12 percent of the world's population; the Aspire is for the other 88 percent.

Huang Miao Hua, an ebullient woman with waist-length hair, shows up and explains that the crew made two Aspire prototypes with a mere $60,000 and three months of sleeping in their classroom. Huang flops down on the shelf in back. What's that for? I ask. "A bike or a rowboat." Why? Because anyone who drives in China can expect to be caught in traffic jams. The rowboat would be for weekend outings.

A concept car for traffic jams! The fact that the Wuhan Institute has made this tiny creature for less than the air freight of most of the other competitors at Bibendum makes it all the more fabulous.

With a target price of $12,000, it's an electric vehicle, so it won't add to the already ponderous smog. Onboard GPS helps identify traffic snarls, allowing our future driver to park (the doors open upward to accommodate tight spaces) and continue on to work with that onboard bicycle. The Linux-based operating system even provides an electronic jack, so that if you do get stuck in traffic, you can simply plug into the CPU and get work done. Instead of an instrument panel, the car has an LCD computer screen, because it's cheaper. And the car has no dashboard, which makes the space inside feel more like a room than a car.

The Aspire's realism is enhanced by an enormous optimism. No sooner do I sit in the passenger seat than I too have the "Wuhan grin." The view out the curving front window is both cute and commanding. Because the electric motor isn't working right, Huang and her students start the car with a push and we're off. The motor kicks in with a horrific high-pitched yowl, as if an animal is being tortured. People stare, then back hesitantly out of our way. Watch out! We'll smile you to death. The engineer shouts above the screech, "It has a few problems. But it has a happy feeling."

At a rally showcasing an idealized future for autos—one of, as GM puts it, "no compromises," the Aspire is *all* compromises. Its slow speed, small size, and agreeable demeanor wouldn't go in Detroit. But it has a clear message: You will sit in traffic jams full of ambitious youngbloods like yourself, but you will have fun, you will be productive, you will prosper, you will spend your weekends rowing a boat. In the United States, cars have always been about getting away from it all; the Aspire is about getting along.

What I really love about sitting in the Aspire is its lack of nostalgia. It doesn't make me miss my grandma's boatlike Cadillac, or even my own pickup truck. I don't feel my hair flying in the breeze or start quoting Kerouac. It's a car so involved in the future, it's lost all attachment to the ideas of the past. "This reminds me of growing up in Japan in the 1960s," says a designer for one of the Big Three a bit wistfully, from Bibendum's sidelines. "Everyone had that look in their eyes. They had a sense of purpose and a determination to push the country forward."

At the end of the challenge, the Aspire wins a special design award. China does surprisingly well, considering it didn't enter a single vehicle in the previous year's Bibendum. This year it entered forty-three vehicles, of which nineteen scored very high in the rankings on emissions and fuel economy. However, not a single Chinese vehicle did well in the acceleration category. Dominating the road doesn't seem to be part of the plan.

When I last see them the Aspire crew is still euphoric, pushing the car through a six-point turn to get it out of the garage. Huang talks about all of the people who gave time and money to make this little egg car and haul it all the way to the contest. "We wanted to do something good for the Chinese automotive industry," she says. "It's good for the country! They gave without expecting any return. It's the Spirit!"

**The Aspire** is part of China's national "863 Project" to research and design electric, hybrid, and hydrogen cars. On a very modest budget of $123 million over five years, Professor Wan Gang is steering the program to create dozens of different competing cars and technology. A compact man in his fifties who retains the pink cheeks and enthusiasm of a boy genius, Wan is a firm believer in the leapfrog.

In 2000 China's Ministry of Science and Technology approached Wan, who had been working in development and strategic planning for Audi in Germany for more than a decade, to return to China and whip the domestic auto industry into shape. Wan took a look at China's hundreds of car manufacturers—with little manufacturing prowess and even less intellectual property—and concluded that it would be impossible to catch up to the West in traditional car technology. Trying to seize ground in new technology would be the country's best bet—but still a long one.

Wan meets me in the visitors' lounge at Tongji University, where he is president. The lounge is huge and done in Deng-era brocade. Huge oversize armchairs hug the walls. We sit down to talk, then he quickly jumps up and moves to another room, full of steel office furniture that could be in Berlin, Singapore, or Los Angeles. Then he lays out the plan for leapfrogging. It's not just about the technology, he explains, it's also about knowing markets and luring in foreign partners. He pulls out a piece of paper. "I'm trying to demonstrate that the picture is reasonable and practical," he says, sketching a grid with categories for electric cars, hybrids, CNG cars, and hydrogen fuel cells.

At first glance the grid appears to contain four distinctly different vehicle types: Electric has been largely abandoned by the big auto makers, hybrid is seen as a helper for internal combustion engines, CNG is based on natural gas, and hydrogen has been seen as an exotic outlier compared to the other technologies on Wan's list.

Wan, however, thinks of hydrogen less as a replacement for gasoline than as a glorified battery. Hydrogen is a way to store energy from various sources—coal, solar, nuclear, or hydro—and eventually convert it to electricity. This places hydrogen and electric vehicles on the same continuum, meaning that any sort of battery or electric power train developed now can be used in fuel cells later. Likewise, Wan sees hybrid technology as largely a question of storing braking energy in a battery.

"Engineers in the States say hybrids are transitional, but I believe the technology will last a long time," he says, drawing arrows across the grid to show how hybrid technology will make hydrogen cars more efficient. Finally, the pièce de résistance: CNG cars require a fueling infrastructure that can be adapted to produce hydrogen.

He holds up the paper covered in arrows, saying "China has an advantage of not being burdened by previous investment, so we can leapfrog the technology forward."

No one knows if hydrogen will eventually be a commercially viable fuel. The road from now to hydrogen is long, treacherous, and expensive. And that, in a way, is the beauty of Wan's drawing—by spreading his resources around among many competing technologies, he's sure to come up with several that are successful, regardless of what happens with hydrogen. And the 863 Project is structured in a way that gets the benefits of both cooperation and competition between rival universities and companies.

Around the country there are dozens of experimental cars, buses, and motorcycles. Some have taken off very fast. In 2000 a few companies began selling electric bicycles. The first year they sold 330,000. By 2005 they were selling 10 million of these cheap ($300) scooters, which don't pollute and use just 15 cents' worth of electricity to go fifty miles.

A similar shoestring aesthetic informs the development of a hydrogen bus at Beijing's Tsinghua University that is designed to be less than half the price of the Mercedes fuel-cell bus and use one-third the hydrogen per kilometer. The bus saves money on labor costs, but also by using some Chinese-designed components, including a Shen Li fuel-cell stack and local hydrogen tanks. Tsinghua's bus goes only 35 miles per hour, because what bus driver on the crowded streets of Beijing or Bombay is ever going to go faster than that? Dr. Ouyang Minggao, the coordinator of the hundred-person bus development team, is counting on the city of Beijing to install hydrogen in its CNG stations so that he can test fifteen buses by 2008, and he sees a niche market for hydrogen buses in developing countries. "China will be very competitive," he says. "Not the best quality, but reasonable."

Wan calls this success on a tiny budget "another kind of leapfrog," though he cheerfully acknowledges that their designs are far less reliable than those of Western companies. The way he sees it, China has something the competition doesn't. "Even though their investment is a hundred times higher, it is difficult for them to form an atmosphere for developing an industry," he says.

If the program succeeds, it'll be on the strength of its ideas and the collective will of its leaders. At age sixteen, during the Cultural Revolution, Wan was sent so far into the countryside that the bus stop was ten miles away. He spent eight years cut off from the world, eventually falling in love with the internal combustion engine while dismantling the tractor belonging to his collective farm. His collective built an entire village—hospital, electrical generator and electrical grid, shops, and roads—from nothing. Now he spins about Shanghai in a chauffeur-driven Audi, a transformation that seems no less miraculous than the idea of starting a hydrogen economy by 2020.

**Still, many people** think that hydrogen is not the way for China to go. The economist Dr. Yang Yiyong, for example, compares it to setting up a colony in space. And Dr. Zhou Dadi, the energy theorist, thinks it is too expensive. The true believers turn out to be in Shanghai, where hydrogen is part of the city's long-term strategy to become the new Detroit, only cleaner.

Downtown Shanghai is a jarring mix of the old and the ostentatious. Near the People's Park, for example, there is a Maserati dealership. I meet Cai Xiaoqing, a former space program technocrat, at a giant downtown hotel with a commanding view of the city. Cai sports an astronaut's brush cut and his foot jiggles anxiously.

Cai works for Shanghai's economic commission as the long-range planner for the city's auto industry. As elsewhere in China, "long range" means *really* long range. Cai is quite comfortable doing math in the billions in his head. He talks about how Shanghai will profit from the 120 million cars on the country's roads in 2020, but he's also got an eye on the rest of the world.

"Of course, since I'm responsible for the development of the car industry, I hope we'll make more and more cars and more people will have cars," he says with a half-grin. "But energy and the environment are very big binding conditions on our development."

Logically, the only way to ensure success, and jobs for the rural im-migrants who continue to throng into the city, is to prepare for what Cai calls the "end future" of hydrogen and alternative fuels.

With the right technology, he thinks, China could make hydrogen from coal or nuclear power, then use that to fuel cars. More important, once China develops hydrogen cars that work, no one will be able to un-dercut the price. Shanghai could own the market for hydrogen cars for decades.

So now Cai looks down at Shanghai's maze of streets and imagines them filled with ten thousand scurrying hydrogen fuel cells, like so many lab rats, by 2015. He bounces slightly as he talks about the city as a labo-ratory, with ten fuel-cell cars in circulation in 2005, a thousand by 2010, and so on. Getting the cars accepted by customers in 2020 is "a long step," he says, which will require government help with the technology, economics, and politics.

Shanghai will need help. And Cai has his eye on General Motors, which has invested over $1 billion in its fuel-cell program in the United States. "If China developed the infrastructure," he says, "of course GM would like to put those cars to use. I think they see China's big market too."


**For over a year,** Tim Vail, who works on marketing and commercial-ization of fuel cells for GM, has been traveling back and forth to China and liking what he finds. Vail's first love was gasoline—he grew up in a Texas oil family and became licensed as a Mercedes mechanic while still in high school. Before he went to law school, he met his future wife by fixing her car's air conditioner. Now Vail looks at Shanghai's natural gas taxi fleet and sees taxi company owners who are ready to experiment. (He says they've already offered to buy a hundred fuel-cell vehicles when the price and technology are right.) He's looked at Shanghai's $1 billion maglev train and seen a city that's ready to spend serious money on tech-nology. He's looked at a coal-based chemical plant in the city and found a source of industrial hydrogen that should last for the next fifteen years. And most of all he sees a culture that's ready to do "social engineering" to get people to adopt fuel cells. To Vail, Shanghai's ridiculously crowded city center, with its gaggles of nouveau riche trying to conspicuously out-consume each other, is a plus. "More than anywhere else, China could say only fuel cells [in the downtown] without a lot of debate. You would see well-heeled people buying fuel cells if they had enhanced rights," he says.

Larry Burns, GM's VP of Research and Planning, has been traveling the United States pitching his team's latest fuel-cell car, the Sequel. The Sequel is fat and luxurious, acceleration is described as exhilarating, and it seems to combine the best of lunar landing modules with a home entertainment center. It even appears to have gills. But after spending more than $1 billion bringing this "no-compromises" machine to the prototype stage, Burns is girding for the long haul—many billions more bringing it into production and marketing it. To do this profitably, Burns needs markets of scale: China-type scale.

Burns says that after 5 to 10 percent market penetration, hydrogen will reach an "inflection point" and sales will begin to take off in an S-curve. Sometime after that, when hydrogen catches on, the whole economy of cars and oil will be upset, putting trillions of dollars into play in the new hydrogen economy. But we're getting ahead of ourselves here.

For the moment, what's needed is a fast track to the inflection point, and that can be done, Burns says, only with government and industry cooperation. The United States lacks aggressive pro-hydrogen policies, and we value so many chattering opinions that we get what Burns calls "frictional losses." He's frustrated. "A government is not just a tax strategy! One of the important roles they play is to create collective will—like Kennedy going to the moon. A government like China has a better chance than maybe a free country like the U.S."

It's surreal to hear a Detroit executive, one of those legendarily cranky champions of free choice and capitalism, championing a coercive market. But strategically, Burns can't afford to let China be the land of lost opportunity. The United States' exclusive dependence on petroleum has made Detroit vulnerable. Burns dryly calls petroleum dependency "not a robust position" for the industry. If China installs the dozens of small nuclear plants it has proposed and then uses the nuclear energy to create hydrogen, will the United States have a counterstrategy? "Do we just watch them explode?" Burns asks, outraged.

I wondered about Burns's assessment of China's program. Was it really about to take off? I called several experts who said no, in one way or another. Some said that China was too late to the technology party, and some said that the country had too little money and too many gas stations already. One of the optimists, though, was Michael J. Brown, a venture capitalist with the Canadian firm Chrysalix, who's been investing in hydrogen for two decades.

Brown has begun investing in China, and he thinks Shanghai's plans are "eminently doable. And if they went balls to the wall, they could do more." As a way of storing energy, hydrogen makes sense for China, he says, because it works with so many different energy sources. But at the moment, that matters less than the fact that Shanghai has enough hydrogen to fuel half a million cars, and "the first country that gets half a million fuel-cell vehicles will rule the world. They'll figure out a way to make fuel cells cheaper than internal combustion engines. Will you tell me how the rest of the world will catch up?" Brown is putting his money on China. "The air in China is filthy. The U.S.'s air is clean, and so the will is just not there."

**A day before** I leave China, Jiading—Shanghai's auto utopia—holds a ceremony to celebrate the founding of the city three years earlier. While a band plays "Remember the Red River Valley," Shanghai dignitaries mill around an unimpressive golf cart–like vehicle. This turns out to be the Spring Light, the latest car produced by Wan Gang's 863 Project. The Spring Light is a hybrid fuel cell with wheels powered by individual electric motors. The Spring Light looks wimpy and decidedly unfun, not to mention its name, which sounds like a margarine brand.

I am disappointed by the little car, but I start a conversation with one of the engineers, who says he's been working with a team of about thirty students and professors around the clock for the past month to get it finished. The price, he says, is $5,000 to $7,000. I have to ask him to do the math for me twice because I don't believe him. "Five thousand dollars," he repeats. It's a delivery vehicle, designed for crowded city streets. My attitude changes: Instead of being a golf cart, the Spring Light is now the cheapest hydrogen car I've ever heard of, by a factor of ten! Now I understand why Cai, Shanghai's auto planner, is standing off to the side, smoking a cigarette and watching the Spring Light proudly out of the corner of his eye. Someday the city of Shanghai may be crawling with these little buggers, all dribbling water from their tailpipes. Thousands and thousands of Spring Lights, running on great engines of collective will, could turn Shanghai into the auto capital of the world.

The crowd is starting to move toward the auditorium, giving the engineer just enough time to mention that the Spring Light is designed to be a flat platform, so that owners can snap on different exteriors, as if their bodies were different outfits. Inside the theater Wan Gang, the leader of the research program, is showing a photo montage of Chinese-built vehicles set to the music of the Beatles' "Drive My Car." The mood is ebullient, and when Wan Gang stands up to speak, he mentions that the city we're all sitting in, Jiading, was no more than an idea three years ago.

Ideas, even more than oil, are fuel. I don't know where Wan Gang's experiment will lead, or if there will ever be hundreds of thousands of Spring Lights clogging the arteries of Shanghai. But I can imagine, someday not too far in the future, sitting in some car like the Aspire, looking at the world in a very different way.

# EPILOGUE

**Regular Unleaded $3.21⁹⁄₁₀**

One evening in mid-2006 I return to Twin Peaks gas station to buy gas. Over the gas pumps, the station's high canopies drop cubes of violet light. Customers scurry between the bright light at the pumps and the fluorescence of the convenience store. As I wait in line, I check to see that the usual comforts are on the shelves: cigarettes, condoms, air fresheners, Snapple, five flavors of corn nuts, energy bars, energy drinks—they're all here. Where the purple candy wizards stood on the counter there is now a small box of Rosa's Fudge.

Everything in the gas station is in its place, right down to the hedges and the orange cones and the security cameras. I spent the last three years traveling 100,000 miles (and burning more than 3,000 gallons of fuel) and watching the oil world change faster than I could record it, but in the gas station, everything is almost the same as it was when I left it. Congress is still fulminating about investigating oil companies for "gouging" while the Senate is (once again) debating drilling in the Arctic Wildlife Refuge. In 2006 the U.S. auto and truck fleets' ability to accelerate from zero to 60 mph improved by a third of a second, but overall fuel economy didn't budge. Since I sat in Roger's tanker truck in 2003, U.S. oil consumption has continued to rise. The only indication of change is on the sign above the station—$3.22 is nearly double the price per gallon that it was three years ago.

What does the price mean? I now see it the way NYMEX oil analyst Tom Bentz does—as stories. Some stories are general: As the rate of demand has grown faster than supply, risks and prices have risen. The balance of power in the oil market has shifted. And some are specific: Perhaps Hugo Chávez is building a house for Yvonne, the woman who's been waiting on the side of the Caracas cliff for twenty-seven years, instead of investing in more oil production capacity. Or, somewhere in the Niger Delta, another promise has been broken.

Ever since the Arab oil embargo, high gas prices have been perceived as a sign of American weakness. This time, one cause is India's and China's rising oil demand—itself partly the result of American influence in globalizing trade. Now a worldwide middle class aspires to own cars, motivated by the American-born belief that car ownership confers existential, economic, and political freedom upon the driver. India and China, once hopelessly burdened by poverty and huge populations, are now zooming up the on-ramp to the great global freeway. And this makes the solution to high prices far more complicated than building a larger Strategic Petroleum Reserve, weakening OPEC (as some American commentators have suggested), or using development money to drill for more oil in unstable countries.

Anyway, $3.22 doesn't begin to account for all the hidden pennies and costs of oil through the supply chain, from pollution to human rights to military expenses in the Gulf. Gas prices do a good job of telling us what today's risks are, but they don't predict the next risks. How stable is Chad? Are American fuel pipelines secure from catastrophe and from terrorists? The price on the pump can't anticipate that. And what is the opportunity cost of not pushing U.S. industry to create more fuel-efficient technology? The sign doesn't say, and while we wait for a clearer message that change is inevitable, American consumers are paying the costs on our credit cards, and avoiding planning for the future.

The one lesson I've learned from writing this book is that there is no such thing as cheap gas. In oil states like Venezuela, Iran, and Nigeria, discount gasoline is the state's bargain with the populace. Remove the subsidies and governments fall. In China, price-controlled gasoline is another form of social compensation, a mechanism of control. And in the United States? I think we've come to expect cheap gas in a sort of "grand bargain" with our government and oil companies. And even when the sign at the gas station tells us otherwise, we persist in hoping that the price will come back down. It's time for us to demand, and make, real changes.

**When I'm** feeling particularly pessimistic, I find comfort in remembering that oil once saved the whales. The original "alternative fuel," oil owes its birth to environmental pressures, technological innovation, dogged investment, and sheer cussedness. By the late 1850s whales were overhunted and getting scarce. Whale oil, the gold standard of indoor lighting, had reached the astronomical price of $2.50 per gallon in 1850s dollars. In labs around the world, chemists tried to come up with new sources of fuel, refining coal, turpentine, and, finally, a patent medicine named "Seneca Oil." Headaches? Worms? Dropsy? A good dose of Seneca Oil, found oozing out of the ground in Titusville, Pennsylvania, was touted as the cure. Refined, Seneca Oil produced kerosene. Kerosene, on its own, was only marginally useful . . . until a glass blower in Austria created a cheap lamp with a glass chimney that neither smoked nor smelled bad.

To use the oil commercially, someone needed to get it out of the ground. Enter Colonel Edwin Drake, a former train conductor with a free railroad pass, who landed in Titusville, Pennsylvania, in 1857. Like some improbable character in a Broadway musical, Drake arrived in a top hat, told charming stories, and set about drilling for oil using the technology of local salt drillers. Drake was many things, but he was not a colonel. He had adopted the name to impress the locals, on the advice of his wealthy backers in the Northeast. Drake's task was so ridiculous the locals called him "Crazy Drake," but he gradually became the man of gumption he impersonated. In early August 1859 his last investor sent a money order telling him to pay his debts and come home. But before Drake received the letter, the drill bit slipped into a fissure in the rock sixty-nine feet deep and oil shot out of the hole—the first gusher.

The Oil Age began in a messy frenzy of greed and speculation, and it spawned technologies ranging from the oil pipeline to the tanker to the auto and the airplane. It created mini-booms everywhere it was found, and in the United States it created a long sustained burst in productivity and wealth that lasted well into the twentieth century. Oil brought other innovations, too. John D. Rockefeller organized the chaotic markets of supply, transport, and demand according to his own design. He limited competition and built refineries, sales networks, brands, and an organizational structure, which became the modern corporation, an entity that now dwarfs many governments and has reorganized the culture and politics of the world. And here, too, the oil industry's greatest critic, journalist Ida Tarbell, grew up on the banks of Oil Creek and later wrote a series of pioneering investigative stories that led to the adoption of the anti-trust laws that broke up Rockefeller's Standard Oil.

**The task** of the next fifty years is to coax many new Drake's wells into existence: We need the chemists, the lamps, the investors, and more than one former train conductor to create the mixtures of technologies, fuels, investments, and policies working in concert. We need many fuels, not just one. As we try to move toward energy that is both economically and environmentally sound, we need to question whether the innovations of Oil City—the cars, the corporations, the antitrust laws, the network of roads, the murky relationship between government and industry—are still working to our advantage. Are they giving us the strategic flexibility we need?

As the United States prepares to move on, metaphorically, to the next whale, we also need to reconsider what energy is. Drake's gusher was in the ground, but the next ones may be in our brains. We are used to thinking of energy as something that can be found in reservoirs and coal seams, but the reserves of the future lie elsewhere—in new forms of energy, but also in efficiency itself—in doing more with less. In fact, we've been burning brains for the past thirty years—three quarters of America's increasing energy needs during the last three decades were met by improving efficiency. The next reserves of efficiency are lying out of sight—just as oil did until Drake found it with a drill bit in 1857—but they are huge. If the United States could increase its energy efficiency by just 3 percent per year between now and 2100 (somewhat less than the 3.4 percent yearly improvements the United States made between 1980 and 1986), the economy would grow dramatically while reducing the world's energy needs to half of their current levels. The implications of this premise are so huge that its proponents, who include Fermi Prize winner Art Rosenfeld, describe it as the "Conservation Bomb."

The technologies and fuels of the future will not come cheaply, easily, or even soon. In the summer of 2005, I was test-driving a big fuel-cell hydrogen truck made by GM (powered by two fuel-cell stacks, it can accelerate from zero to 60 mph in 19 seconds, and it's under contract to the military) when its power supply conked out. A siren began to wail and the light labeled "safety line" flashed as I parked the truck on a burned-out city block laced with razor wire. The two engineers in the truck peered at their computer screens. One yelled, "Can you get me a DOS prompt?" while the other called for a wrecker. Eventually, they solved the problem by rebooting, clinching my suspicion that what appeared to be a military truck was at the same stage of development as word-processing software was in the mid-1980s. I finally understood what engineers mean when they say that commercial hydrogen is at least fifteen years away—and that is assuming that the many problems inherent in hydrogen supply, production, and vehicles work out at all.

In the meantime, the United States will become increasingly dependent on foreign supplies of oil, particularly from the Middle East and West Africa, and we'll need to manage these relationships with more care and commitment than we have in the past. Oil diplomacy, long outsourced to oil companies, and increasingly to the U.S. military, needs attention and leadership. The special relationships the United States nurtured with countries like Venezuela and the security guarantees offered to Saudi Arabia have lost their appeal; and the threats, which include sanctions and military intervention, have lost their effect. The problems of the villages at the wellheads must be solved—if not for moral reasons, then for self-interested, economic ones.

The United States could reimagine the bargain, offering environmental and economic security (particularly jobs and non-oil investment) to the citizens of oil-producing countries while imposing a tax on oil imported into the United States to pay for the program. Where the United States now champions democracy at the voting booth, we should consider supporting governments with a system of taxes and more responsive leadership. All of this will require American leaders with the guts to abandon the status quo and look toward the future with a critical and strategic eye.

There is no reason to imagine that the only ending to the oil story is a frantic and brutal scramble for resources dominated by an increasingly craven "addicted" United States. If the United States chooses, it could articulate a new American Dream guaranteeing environmental and economic security along with prosperity and personal freedom. The United States could put its considerable money and political will into creating new kinds of vehicles and fuels, while creating incentives to use fossil fuels more efficiently, unleashing a double whammy of new fuel and the "Conservation Bomb." Just as Drake's well started the huge upheaval in technology and politics that led the oil age to foster globalization, the next grand vision has the potential to bring with it technology, profits, security, and benefits we can't even imagine yet.

You won't find any of these ideas at your local gas station amid the Snapples, two Sominex and a folded paper cup, and the hidden pennies. The next trip, like all American trips, starts at the gas station . . . and then leaves it behind.

# ACKNOWLEDGMENTS

This book was a long and fortunate collaboration with many people in many countries.

There would be no book at all if it weren't for the many people in and around the oil industry who let me into their lives. Among them are Joe Mullin, Ian Buist, and David Dickins, who first introduced me to oil in Prudhoe Bay. Michael Gharib and B. J. Singh of Twin Peaks Oil warmly let me hang around the gas station. Jerry Cummings, Herb Richards, Mark Mitchell, Roger, Chris, and the people at Robinson and Coast Oil invited me into their complicated business. Walter Neil, Ken Cole, and the people at BP Carson brought me inside their beast. C. D. Roper generously let me into his mud-logging trailer and his life for an awfully long time. R. L. Gaston's thoughtful words and Texas mug stuck with me. Drew Malcolm of the Department of Energy and the many people at the SPR. Tom Bentz, Ed, Jaime, and the people at the NYMEX.

In Venezuela, Jorge Hinestroza, Charlie Hardy, and Suzana were amazing guides, as were the people of the new and old PDVSA and the Gente Del Petroleo. In Chad, Ian Gary, Nicole Poirrer, and Akibou Djounouma helped me get my bearings in a strange place. Jean the carver introduced me to the people of Kome Atan. In Iran, Aresu Eqbali is the original lady of the stuffed shark. The people of NIOC were very generous. Jahan Aliyeva's tour of Azerbaijan was unforgettable. In Nigeria, Dennis Okatubo took me under his wing and Von Kemedi made it possible for me to stay for long enough to settle into Nigeria. The people of the kingdom of Alcassa, Bill Knight, and Tracey Draper made it possible for me to visit Alcassa. Rex Chen and Alysha Webb were among the many people who helped me see New China.

I'm grateful to the many people who've offered to read chapters and made suggestions: Scherle Schwinninger, Marc Herman, Connie Voisine, Louise Steinman, Doug McGray, Terry Lynn Karl, Mike Ceaser, Ike Okonta, Rose George, Shannon Brownlee, Kerry Tremain, Wienke Tax, and David Buchbinder. And special thanks to Leah Mezzio for corraling my buckets of sources into organized notes. Bill Van Parys, Laura Moorhead, Burkhard Bilger, Katie McColl, and Stephan Ruiz assigned stories that helped keep this project rolling along.

The Drake Well Museum of Titusville, Pennsylvania, provided the cover of the song "Oil on the Brain," which appears in the front of this book. If you have read this far in the acknowledgments, you should consider a trip to the museum or nearby Oil City on your next vacation.

The New America Foundation's support has been key in the creation and completion of this project.

I was lucky to have the support and encouragement of my parents, David and Susan Margonelli. Mary Roach first suggested that I write a travel book on oil when I was rattling on about petroleum one day. Jay Mandel sheep-dogged the idea into a proposal. Coates Bateman sent a constant stream of encouraging e-mails to the weirdest places. And Lorna Owen edited patiently as it arrived. Also helpful, but in harder to define ways, were Zohreh Soleimani, Skip Laitner, Dudley Althaus, and Peter Barnes and Mesa Refuge.

In Berkeley, Matt, Janna, Brendan, Caitlin, and Jackson the dog were excited when I came home and understanding when I was away. Kitty, Dave, Lucy, and the increasingly big Little fed me a lot of meals. Joe Loya, Beth Lisick, and the people at the Grotto graciously listened to me fret. Michael and Alonza Lasher introduced me to the wonders of the chard pancake and Mango Queendom.

# NOTES AND REFERENCES

All quotes in this book were obtained in interviews I did between 2003 and 2006, except where otherwise noted. In some cases names were omitted for easier reading, and in others because the subject requested that he or she remain anonymous. Facts cited in the text are referenced in the following notes, but some citations have been compressed. If you have questions about sources, please contact me at margonelli@yahoo.com.

*CHAPTER ONE*

8  *194 million licensed American drivers*: Stacy C. Davis and Susan W. Diegel, *Transportation Energy Data Book*, 24th ed. (Oak Ridge, Oak Ridge National Laboratory, 2004), table 8-2.

8  *$4.4 billion in salty snacks; purchase of $25 billion in lottery tickets; $323 million on cold medicine in gas stations in 2001*: National Association of Convenience Store's *State of the Industry Report*, July 2005. Statistics for 2001 are from the 2002 501 report. (Alexandria, VA, 2002), p. 46.

8  *gasoline vapors in California stations total 15,811 gallons a day*: Communication from Dmitri Stanich of the California Air Resources Board based on information from the Reactive Organic Gases Projected Emission Inventory Database 2004 for California.

8  *Four out of five people feel, on a gut level*: Thomas Turrentine and Kenneth Kurani, "The Household Market for Electric Vehicles: Testing the Hybrid Household Hypothesis—A Reflexively Designed Survey of New-Car-Buying, Multi-Vehicle California Households," Institute of Transportation Studies, University of California, Davis, May 15, 1995, pp. 92–93.

8  *Japanese auto executives have hired American anthropologists*: Interview by author with Steve Barnett, president of the consumer research and insights group at Smart Revenue, 2003.

8  *1,143 gallons per household per year*: Energy Information Administration, *Household Vehicles Energy Use: Latest Data and Trends* (Washington, DC, Nov. 2005), U.S. Per Household Vehicle-Miles Traveled, Vehicle Fuel Consumption and Expenditures, p. 57.

8  *140 billion gallons of gasoline*: Energy Information Administration, Country Analysis Brief, updated November 2005.
http://www.iea.doe.gov/emue/cabs/Usa/Oil.html

9  *168,987 gas stations*: National Petroleum News, 2005 Station Count.

10  *$2,141 a year to gasoline theft*: National Association of Convenience Stores, "Gasoline Theft at Convenience Stores," July 2005 fact sheet.

11  *Nearly nine percent of U.S. robberies*: Rosemary J. Erickson and Sandra J. Erickson, "Summary and Interpretation of Crime in the United States 2004 *Uniform Crime Report*, Federal Bureau of Investigation, released October 17, 2005" (Athena Research Corporation, Nov. 2005), p. 4.

11  *Urban legends about AIDS-tainted hypodermic needles*: Barbara Mikkelson and David P. Mikkleson, Urban Legends Reference Pages, 1995 to 2006, www.snopes.com/horrors/mayhem/gaspump.asp.

11  *a customer driving a Ford 150 experienced . . . a customer driving a Honda Accord*: Robert N. Rennkes, "Fires at Refueling Sites that Appear to Be Static Related," Petroleum Equipment Institute, 2006.

11  *"pump rage"*: Douglas Belkin, "High Prices Embolden Gas Bandits," *Boston Globe*, July 21, 2005.

11  *After 9/11 people . . . attacked one hundred 7/11 clerks*: Richard Louv, "Security Experts Take Violence Against Arab-Americans Seriously," *San Diego Tribune*, October 14, 2001.

11  *a list of tips to discourage customers from attacking employees*: National Association of Convenience Stores, "Helping Store Employees Deal with Ethnic Hostility," Press Release, August 2002.

13  *sunglasses have a 100 percent markup . . .*: National Association of Convenience Stores, "2002 *State of the Industry*," p. 43.

14  *$132 billion spent in convenience stores*: National Association of Convenience Stores, "Convenience Store Industry Sales Hit New Highs in 2004," Press Release, April 12, 2005.

14  *one in six gas stations has closed . . . ; independents own about 35 percent of the stations . . . ; the independents all use a generic formula . . . ; and discount the wholesale gas; wholesale gas can be 10 cents more . . .* : George Anders, "As Oil Prices Swing, Gas-Station Owners Try Futures Market," *Wall Street Journal*, July 21, 2005.

17  *An investigation by the Federal Trade Commission in 2000*: Elizabeth Douglass and Gary Cohn, "Zones of Contention in Gasoline Pricing," *Los Angeles Times*, June 19, 2005. Justine Hastings, Prepared Statement before the California State Assembly Select Committee on Gasoline Competition, Marketing, and Pricing, April 28, 2004, p. 10; aida.econ.yale.edu/~jh529/JHastings_Testimony_CAState-Assembly_042804.pdf.

17  *forcing gas stations to sell with a minimum markup*: Federal Trade Commission, *Gasoline Price Changes: The Dynamic of Supply, Demand, and Competition* (Washington, DC, FTC, 2005).

18  *"merchandise shrink" costs*: National Association of Convenience Stores, "2002 State of the Industry," p. 72.

19  *Numbers of gas stations in different decades*: John A. Jakle and Keith A. Sculle, *The Gas Station in America* (Baltimore: John Hopkins University Press, 1994), chapter 3.

19  *Shell threw up 100 identical gas stations between . . . ; Frank Urich . . . opened a station . . .* : John Margolies, *Pump and Circumstance: The Glory Days of the Gas Station*, (Boston: Little, Brown, 1993), p. 112.

19  *Frank Lloyd Wright saw stations . . . ; "How many trusties and lusties . . .* : Daniel I. Vieyra, *Fill 'er Up* (New York: Macmillian, 1979), p. 63.

20  *the oil industry dropped $150 million on trading stamps*: John Jakle and Keith Sculle, *The Gas Station in America*, p. 71.

22 *letter from Hugh Lacy, Urich Oil Company's senior vice president*: Fred C. Allvine and James M. Patterson, *Competition Ltd: The Marketing of Gasoline* (Bloomington: Indiana University Press, 1972), pp. 291–294.

23 *Fuel economy actually went down*: Energy Information Administration, *Annual Energy Review 2004*, Table 2.8.

24 *"Obscene profits"*: Daniel Yergin, *The Prize: The Epic Quest for Oil, Money & Power* (New York: Simon & Schuster, 1991), p. 658.

24 *gas went from 63 cents a gallon . . .* : Energy Information Administration, *Annual Energy Review 2004*, Table 5.24.

24 *In 1980 the average passenger car*: Ibid., Table 5.25.

25 *"ripping off the public"*: Humphrey Taylor, "Public Perceptions of Airline and Oil Industries Plummet," *The Harris Poll #24*; May 24, 2001; http://www.harrisinteractive .com/harris_poll/pinterfriend/index.asp?PID=238 (Accessed 10/15/03).

25 *By 2003 only 4 percent believed . . .* : Humphrey Taylor, "Attitudes to Government Regulation Vary Greatly for Different Industries," *The Harris Poll #19*, April 2, 2003.

25 *"Belief in oil industry conspiracy . . .* : Eric R.A.N. Smith, "Public Attitudes Toward Oil and Gas Drilling Among Californians: Support, Risk Perceptions, Trust, and Nimbyism: Final Technical Report," U.S. Department of the Interior, University of California, Santa Barbara, 2005, p. 53.

26 *congressional Minority leader Nancy Pelosi . . .* : Marc Sandalow, "Drive Less? Politicians Won't Ask," *San Francisco Chronicle*, April 28, 2006.

28 *Number of Americans who'd switch stations for price differences*: National Association of Convenience Stores, *State of the Industry Report 2002*, p. 67.

30 *Between 2003 and 2005, the price of gasoline increased*: Gas price rise information from AAA Fuel Gauge Report; "Large Majorities of US Adults Expect Gas and Heating Prices to Rise," *Harris Poll #44*, May 25, 2005, which included polling data from 1,160 adults surveyed in 2004 and 2005 who said their gasoline spending per month increased $94 between 2003 and 2004 and then another $84 between 2004 and 2005, for a total increase of $178 per month or $2,130 per year, or a price increase of roughly $3 per gallon on 700 gallons.

30 *The average American household spends*: U.S. Department of Energy, *Transportation Energy Data Book* (Oak Ridge, TN: U.S. Department of Energy), 2004 Table 8.3.

30 *In 2005 the Automobile Club of America*: American Automobile Association, *Your Driving Costs 2005*; www.aaawa.com/news_safety/pdf/Driving_Costs_ 2005.pdf

30 *a tiny 4 percent of car buyers*: Beth Rusert and Jennifer Larsen, "SUV and Truck Owners Rarely Consider Gas Mileage When Purchasing Vehicles," Maritz Automotive Research, December 2, 2004.

31 *"I earn more money by presenting myself as successful"; "engage in a type of limited economic rationality"*: Kenneth S. Kurani and Thomas S. Turrentine, "Automobile Buyer Decisions about Fuel Economy and Fuel Efficiency: Final Report to the United States Department of Energy and Energy Foundation," University of California at Davis, September 1, 2004, p. 27.

31 *if gas prices stayed at $4*: Austan Goolsbee, "The Grip of Gas: Why You'll Pay through the Nose to Keep Driving," Slate.com, September 27, 2005.

31 *Some out-of-state gamblers*: Rod Smith, "Gas Prices Could Hurt LV Gaming," *Las Vegas Review-Journal*, October 4, 2005.

32 *People who live in crowded areas*: Texas Transportation Institute, *The 2005 Urban Mobility Report* (College Station: Texas Transportation Institute, 2005).

32 *People actually prefer a half-hour commute*: Alan Sipress, "Not All Commuters Driven Crazy; Many Motorists Say They Like Solitary Time Behind the Wheel," *Washington Post*, October 18, 1999.

CHAPTER TWO

34 *Daily, men like Roger make 50,000 deliveries*: Office of Senator Charles Schumer, "New FBI Warning about Truck Bombs at High-Profile NYC Sites Shows Need for Comprehensive Anti-Terror Truck-Bomb Plan," Press Release, May 30, 2004.

34 *Only about 1 in 1,000 of those accidents*: U.S. Department of Transportation, "Hazardous Material Safety," Hazardous Materials Information System Summary by Class. Statistics available on the Web, aggregated for years 1999–2004, for incidents involving "Flammable-Combustible Liquid."

35 *the gasoline releases the same amount of energy as 194 tons of TNT*: Richard A. Muller, "Cropduster Terrorism," *Technology Review Online*, March 11, 2002.

35 *When a double trailer accidentally overturned near the Pentagon*: Tom Jackman, "Gas Truck Driver Dies in Fiery Va. Wreck," *Washington Post*, December 23, 2004.

35 *adding an extra 65 gallons of fuel consumption*: Thomas F. Golob and David Brownstone, *The Impact of Residential Density on Vehicle Usage and Energy Consumption* (Berkeley: University of California Energy Institute, 2005), p. 22.

35 *traffic jams . . . 2.3 billion gallons*: David Shrank, *The 2005 Urban Mobility Report* (College Station: Texas A&M University System, Texas Transportation Institute, 2005).

35 *as bigger trucks and cars*: Transportation Energy Data Book, pp. 2–3.

37 *In 2004 state governments collected $35 billion*: Federal Highway Administration, "State Motor-Fuel Taxes and Related Receipts of 2004," U.S. Department of Transportation; http://www.fhwa.dot.gov/policy/ohim/hs04/htm/mf1.htm

37 *The impact of gas taxes*: David L. Greene, Testimony to the U.S. House of Representatives Science Committee, "Improving the Nation's Energy Security: Can Cars and Trucks Be Made More Fuel Efficient?" February 9, 2005, pp. 4–5.

39 *Gasoline consists of 150 chemicals*: Agency for Toxic Substances and Disease Registry, "ToxFAQs for Automotive Gasoline," Atlanta, GA: U.S. Department of Health and Human Services, November, 2004. And ATSDR: "Benzene Toxicity and Physiologic Effects."

39 *Verma's research suggests that Roger*: Dave Verma, et al. "A Simultaneous Job-and-Task-Based Exposure Evaluation of the Petroleum Tanker Drivers to Benzene and Total Hydrocarbons," *Journal of Occupational and Environmental Hygiene* (November 2004).

43 *One study based on a 2003 supply shortage in Phoenix*: Federal Trade Commission, *Gasoline Price Changes: The Dynamic of Supply, Demand, and Competition* (Washington, DC, 2005), pp. ii–iii.

43 *. . . Arizona's governor set up an 800 number*: Arizona Office of the Governor, Kinder Morgan Pipeline News Conference, August 19, 2003.

44 *The Energy Information Agency*: Reuters, "U.S. Petroleum Demand to Grow 37 Percent by 2025—EIA," December 10, 2004.

45 *161,000 miles of fuel pipelines in the United States*: Office of Pipeline Safety, "Liquid Pipeline Total National Mileage," 2005, http://ops.dot.gov/stats/lpo.htm.

45 *Etkin's data shows*: Dagmar Schmidt Etkin, Environmental Research Consulting, personal communication with the author, April 2006.

46 *A 1999 pipeline break in Bellingham*: Paul Shukovsky, "Criminal Indictments in Deadly Pipeline Explosion," *Seattle Post Intelligencer*, September 14, 2001.

46 *A report on pipeline safety*: GAO, "Pipeline Safety: Management of the Office of Pipeline Safety's Enforcement Program Needs Further Strengthening," GAO-04-801, July 23, 2004, p. 7.

CHAPTER THREE

48 *If all goes well, by this time tomorrow morning*: BP Carson refinery fact sheet March 26, 2006, divided by minutes.

48 *according to the latest EPA numbers*: Mark Merchant, "EPA Releases Latest Data on Toxic Chemicals Nationwide," Environmental Protection Agency, Region 9 press release, April 12, 2006, p. 3. BP Carson 556,000 pounds for 2004 Toxic Release Inventory, divided by 365 days. The Toxic Release Inventory includes things like offsite waste transfers as well as air emissions. The specific toxics released in 2002 include (in order of weight) ammonia, methanol, toluene, xylenes, benzene, hexane, and 1,3 butadiene.

49 *Without admitting guilt, the company agreed to pay*: Press releases from Air Quality Management District: "AQMD Seeks $319 Million from BP for Air Pollution Violations," March 13, 2003; "AQMD Files $183 Million Lawsuit Against BP for Air Pollution Violations," January 20, 2005; "AQMD and BP Settle Refinery Emission Violations for $25 Million in Civil Penalties, $6 Million in Past Fees, and $50 Million for Community and Clean Air Projects," March 17, 2005; and interviews by the author with Pang Mueller of the AQMD.

50 *Crude arrives as a stew of hydrocarbon chains*: Michael Freemantle, "What's That Stuff?" *Chemical and Engineering News* 77, no. 47 (November 1999): 81

50 *A refinery sorts these molecules as*: Chevron Corporation, *Motor Gasolines Technical Review*, Chapter 3, p. 27; www.chevron.com/products/prodserv/fuels/bulletin/motorgas/.

51 *BP plans to revamp this process*: Elizabeth Douglas, "California: Plan Unveiled for Hydrogen Power Plant," *LA Times*, February 11, 2006.

52 *in March 2005, a catastrophic explosion and fire*: BP Annual Review 2005, "Making Energy More Reliable," p. 30.

57 *And in 1997 the U.S. EPA did an audit*: U.S. Environmental Protection Agency, "Audit of Region 9's Administration of the California Air Compliance and Enforcement Program," Western Audit Division, San Francisco, CA, July 24, 1997.

57 *In early 2006 the EPA realized*: Environmental Protection Agency Evaluation Report, "EPA Can Improve Emissions Factors Development and Management," Report No. 2006-P-00017, March 22, 2006, pp. 11, 12.

58 *between 1999 and 2003, 83 percent of the gases vented*: South Coast Air Quality Management District, Evaluation Report on Emissions from Flaring Operations at Refineries, Version 1, September 3, 2004.

58 *One California refinery made back its investment*: Ernst Worrell and Christina Galitsky, "Profile of the Petroleum Refining Industry in California: California Industries for the Future," Berkeley: Lawrence Berkeley Laboratory, March 2004, p. 41.

59 *To go a single mile*: "EPA Emissions Facts," from the offices of Transportation and Air Quality, April 2000.

60 *about 70 percent of the excess risk*: All figures in this paragraph are from: Rachel Morello-Frosch et al., "Environmental Justice and Regional Inequality in Southern California: Implications for Future Research," *Environmental Health Perspectives*, 110 (Suppl. 2), (April 2002): 149–154.

60 *for children, living within 250 feet of a major roadway raises*: Rob McConnell et al. "Traffic, Susceptibility and Childhood Asthma," *Environmental Health Perspectives* 114, no. 5 (May 2006).

61 *Studies by the EPA*: Environmental Protection Agency, "New Source Review: Report to the President," June 2002, p. 1, accessed at http://www.epa.gov/nsr/publications.html.

61 *In 2005 BP's upstream profits were six times its downstream ones*: BP Annual Review 2005, p. 28.

61 *"If consumers' demand is not sensitive to price . . ."*: Severin Borenstein, James Bushnell, and Matthew Lewis, "Market Power in California's Gasoline Market," (University of California Energy Institute, Center for the Study of Energy Markets, Paper CSEMWP-132, May 2, 2004, p. 13.

62 *BP's first-quarter trading update*: BP PLC "BP First Quarter 2006 Trading Update," Press Release dated April 5, 2006, available at http://www.bp.com/press.

62 *Globally, refineries were using*: Thomas O'Connor, ICF Consulting, testimony before the House Government Reform Committee, October 19, 2005, Exhibit 5.

64 *The amount of oil products*: Energy Information Agency, *Annual Energy Review 2004*, Refinery Input and Output, Selected Years, 1949–2004, Table 5.8.

CHAPTER FOUR

66 *According to the weekly rig count*: Total number of gas rigs in the United States: Baker Hughes Rig Count, August 8, 2003; number of rigs operating in Freestone Country; figures from Texas Railroad Commission.

66 *"Black Giant"*: Roger M. Olien and Diana Davids Olien, *Oil in Texas: The Gusher Age 1895–1945* (Austin: University of Texas Press, 2002), p. 171.

66 *Before a year was out*: James Anthony Clark and Michel T. Halbouty, *The Last Boom* (New York: Random House, 1972), pp. 141, 126.

67 *90 percent of all U.S. drilling rigs*: Federal Reserve Bank of Dallas, Houston Branch, "Houston Business—A Perspective on Houston Economy," 2005.

67 *90 percent of wells*: Kenneth S. Deffeyes, *Hubbert's Peak: The Impending World Oil Shortage* (Princeton, NJ: Princeton University Press, 2001), p. 67.

71 *the Woodbine Sand*: Olien and Olien, *Oil in Texas*, p. 170.

74 *By the end 2003*: International Association of Drilling Contractors, "Summary of Occupational Incidents: US Land Totals," June 3, 2004, pp. 3, 6, 7.

76 *In 2003 one company drilling for gas*: Peggy Williams, "The East Texas Basin," *Oil and Gas Investor*, Houston (February 2004): 7.

77 *In 2004 oil and gas drilling had affected*: Texas Water Development Board, "Aquifers of the Gulf Coast of Texas, Report 365," February 2006, p. 258.

77 *penalties for leaks*: Rusty Middleton, "What Lies Beneath: The Threat from Oilfield Waste Injection Wells," *The Texas Observer*, May 19, 2006.

77 *After the oil output of Texas peaked; Now only one hole out of nine*: Texas Comptroller of Public Accounts, *Rural Texas in Transition* (Austin: Author, 2001), p. 12. http://www.window.state.tx.us/specialrpt/rural/index.html

78 *the company XTO claims; he was attracted to the "romance"*: Hattie Bryant, producer, "Smaller Can Be Smarter, Then Faster—Better," the Small Business School series, episode 1311, Spring 2006, shown on PBS stations.

78 *presentation to potential investors*: XTO Energy Investor presentation, March 2003.

78 *CEO Bob Simpson*: Mitchell Schnurman, "30 Percent Pay Cut Isn't All Bad for XTO Chief," *Star-Telegram*, April 20, 2005.

79 *approximately 89 metric tons of green scum*: Jeffrey S. Dukes, "Burning Buried Sunshine: Human Consumption of Ancient Solar Energy," Department of Biology, University of Utah, August 26, 2002.

79 *marine sediments made of phytoplankton corpses*: Gene Carl Feldman, "Monitoring the Earth from Space with Sea WiFS," found at http://oceancolor.gsfc.nasa.gov/SeaWiFS/TEACHERS/sanctuary_4.html.

79 *a common misconception that oil and gas come from dinosaurs*: M. D. Lewan, "Petrographic Study of Primary Petroleum Migration in the Woodford Shale and Related Rock Units," in *Migration of Hydrocarbons in Sedimentary Basins*, ed. B. Doligez (Paris: Editions Technip, 1987), pp. 113–130, and Donald L. Gautier et al., *The Future of Energy Gases* (Denver, CO: U.S. Department of the Interior, 1993).

84 *The Bossier are the remains*; *nuisance gas*: Rob Karlewicz and Scott L. Montgomery, "Bossier Play Has Room to Grow, Possible Limits in East Texas," *Oil and Gas Journal*, January 29, 2001. pp. 36–43.

85 *A 1995 survey of gas reserves didn't even list Bossier*: National Academy of Sciences, "Summary of a Workshop in U.S. Natural Gas Demand, Supply, and Technology: Looking Toward the Future" (Washington, DC: Author, 2001–03).

86 *Freestone courthouse photocopy machine's annual revenue*: Susan Warren, "As Energy Booms, 'Landmen' of Texas Enjoy a Gusher," *Wall Street Journal*, March 7, 2005.

86 *a long history of cows*: Paul R. Epstein and Jesse Selber, "Oil: A Life Cycle Analysis of Its Health and Environmental Impacts," Center for Health and the Global Environment, Harvard Medical School, 2002, p. 11.

86 *In 2002 Freestone residents declared*: Transactional Records Access Clearing House, "Income Reported on Federal Tax Returns Filed in 2002 Freestone County, Texas," Syracuse University, NY, 2002.

87 *The International Energy Agency estimates*: International Energy Agency, Fact Sheet, "Resources to Reserves: Oil and Gas Technologies for the Energy Markets of the Future" (Paris: International Energy Agency, 2005).

87 *oil and gas technology research*: National Academy of Sciences, *Energy Research at DOE: Was it Worth It? Energy Efficiency and Fossil Energy Research 1978 to 2000* (Washington, DC: The National Academy of Sciences, 2000–1), Table F-34, pp. 212, 213.

88 *In 2005 the Bush administration*: Congressional Budget Office, "279-01— Discretionary: Eliminate the Department of Energy's Applied Research for Fossil Fuels," Washington, DC, 2005, and American Association for the Advancement of Science, "Trends in the DOE Budget, FY 1998–2007," Washington, DC, 2006.

88 *Texas gives $3.5 billion in tax breaks*: Speech by Mark A. Baxter, "Texas Energy Planning Council: An Update of the Economic Impact of Oil and Gas Incentives for the State of Texas," Austin, Texas, April 27, 2004.

88 *all U.S. state incentives for oil and gas exploration; "It behooves everyone"*: Interstate Oil and Gas Compact Commission, "Making a Wise Investment: The Economic Impact and Gas Incentives," Oklahoma City, 2004, pp. 3, 4.

*88  give companies access to $65 billion*: Edmund L. Andrews, "U.S. Has Royalty Plan to Give Windfall to Oil Companies," *New York Times*, February 14, 2006.

*88  lawsuits filed in Colorado*: Kim McGuire, "Tycoon Piloting Royalties Lawsuits," *Denver Post*, February 6, 2006.

*90  pay averages $13 dollars an hour*: U.S. Department of Labor Bureau of Statistics, "Oil and Gas Extraction," Washington, DC, Office of Employment Projections, February 27, 2004.

*91  70 percent of U.S. oil workers; nearly half the industry is between fifty and sixty years old*: Simon Romero and Jad Mouawad, "Holes in the Pipeline; Oil Industry Is Cash Rich but Lacks Machinery and Workers," *New York Times*, October 28, 2005.

*92  A trip to a Kilgore latrine*: Statistics from Clark and Halbouty, *The Last Boom*, pp. 140, 287.

*92  each barrel cost 80 cents; Secretary of the Interior Harold Ickes*: Daniel Yergin, *The Prize: The Epic Quest for Oil, Money, and Power* (New York: Simon and Schuster, 1991), pp. 250, 254.

*97  In the Persian Gulf, the lifting price*: EIA *International Energy Outlook 2005*, p. 31; Darbonne Nissa, "Trash or Treasure" *Oil and Gas Investor*, July 1, 2005; David L. Greene, Donald W. Jones, and Paul N. Leiby, "The Outlook for US Oil Dependence," *Energy Policy*, 26, no. 1 (1997): 58.

*Other Sources Used*

Bank, Gregory C., and Vello A. Kuuskraa, "Gas From Tight Sands, Shales a Growing Share of US Supply," *Oil and Gas Journal*, December 8, 2003.

Haines, Leslie. "Unlocking Tight-Gas Supplies," *Oil and Gas Investor* (March 2005).

Presley, James. *A Saga of Wealth: The Rise of the Texas Oilmen* (New York: G. P. Putnam's, 1978).

Rundell, Walter Jr. *Early Texas Oil: A Photographic History, 1866–1936* (College Station: Texas A&M University Press, 1977).

Selley, Richard C. *Elements of Petroleum Geology*, 2nd ed. (San Diego: Academic Press, 1998).

*CHAPTER FIVE*

*103  Some postwar theorists even supported importing*: Daniel Yergin, *The Prize: The Epic Quest for Oil, Money, and Power* (New York: Simon and Schuster, 1991), p. 428.

*103  Between 1960 and 1973, oil production*: Energy Information Administration, "Table 11.5 World Crude Oil Production," *Annual Energy Review* (2004); www.eia.doe.gov/emen/aer/txt/ptb1105.html.

*103  government allocation programs worsened the shortfall:* "Still Holding Customers Over a Barrel," *The Economist*, October 23, 2003, p. 7.

*104  The British were horrified to hear*: Owen Bowcott, "UK Feared Americans Would Invade Gulf During 1973 Oil Crisis," *The Guardian*, January 1, 2004.

*104  Nixon's government*: Jay E. Hakes, "Administrator's Message, 25th Anniversary of the Energy Crisis of 1973," Energy Information Administration, September 3, 1998.

*104  11 percent of workers in manufacturing positions*: Donald W. Jones, Paul N. Leiby, and Inja K. Paik, *Oil Price Shocks and the Macroeconomy: What Has Been*

*Learned Since 1996* (Oak Ridge, TN: Environmental Sciences Division Oak Ridge National Laboratory, 2002), p. 9.

*104* *"an illusion of U.S. impotence"*: Comptroller General of the United States, "More Attention Should Be Paid to Making the US Less Vulnerable to Foreign Oil Price and Supply Decisions: Report to Congress," January 3, 1978 (Washington, DC, General Accounting Office), p. i.

*104* *"provide credible evidence"*: Committee on the Interior and Insular Affairs United States Senate, *Strategic Petroleum Reserve Plan* (Washington, DC: US Government Printing Office, 1977), p. 4.

*104* **To this day conservatives from the Heritage Foundation**: Charli E. Coon and James Phillips, "Strengthening National Energy Security by Reducing Dependence on Imported Oil," Heritage Foundation, April 24, 2002, p. 10.

*104* *oil be stored in old tanker flotillas*: David Leo Weimer, *The Strategic Petroleum Reserve: Planning, Implementing, and Analysis* (Westport, CT: Greenwood Press, 1982), p. 36.

*104* *put it in large rubber bags*: Committee on the Interior, *Strategic Petroleum Reserve 104*, p. 70.

*105* *"If the federal government is going to pour our money down a rat hole . . ."*: Weimer, *The Strategic Petroleum Reserve*, p. 50.

*105* *the United States dumped more than $37 billion*: Robert Bamberger, IB87050: Strategic Petroleum Reserve CRS Issue Brief for Congress, Washington, DC: National Council for Science and the Environment, 2001).

*105* *Carter said . . . Americans "deeply resented" . . .* : Daniel Yergin, *The Prize* (New York: Simon & Schuster, 1992), p. 662.

*105* *reduced Americans' per capita oil consumption*: Energy Information Administration, "25th Anniversary of the 1973 Oil Embargo" (Washington, DC, 1998), slide 8. http://www.eia.doe.gov/emeu/25opbc/sld008.htm

*106* *the (simulated) price of oil went to $160 a barrel*: Kevin G. Hall, "Simulated Oil Meltdown Shows US Economy's Vulnerability," Knight Review Newspapers, June 24, 2005.

*108* **In late 2002 the price of oil began to rise dramatically**: Minority Staff of the Permanent Subcommittee on Investigations of the Committee on Governmental Affairs, United States Senate, *US Strategic Petroleum Reserve: Recent Policy Has Increased Costs to Consumers but Not Overall US Energy Security* (Washington, DC, 2003), p. 16.

*108* *inflating oil prices by 25 percent*: Senator Carl Levin, "Levin and Collins Urge Suspensions of SPR Oil Deposits to Lower Oil Prices and Increase Private Sector Supplies," News release, February 13, 2004, p. 1.

*112* **In core samples, the salt crystals glitter icily**: James T. Neal, Thomas R. Magorian, and Saddam Ahmad, *Strategic Petroleum Reserve (SPR) Additional Geologic Site Characterization Studies Bryan Mound Salt Dome, Texas* (Albuquerque: Sandia National Laboratories, 1994), pp. 1–63.

*114* **In 1997 police arrested members of the KKK; In 1999, Vancouver police arrested a man; By 2002, pipeline companies**: Paul W. Parfomak, "Pipeline Security: An Overview of Federal Activities and Current Policy Issues," CRS Report for Congress, February 5, 2004, p. 27.

*114* **In the early 1980s a report on U.S. energy security**: Amory Lovins and Hunter Lovins, "The Fragility of Domestic Energy," *The Atlantic Monthly* (November 1983): 20.

115 *Reducing the highway speed limit to around 50 mph*: International Energy Agency, "Saving Oil in a Hurry" (Paris: Imperial College, 2005), p. 116.

116 *and a hispid pocket mouse*: Jack D. Tyler, "Vertebrate Prey of the Loggerhead Shrike in Oklahoma," *Proceedings of the Oklahoma Academy of Sciences* 71 (1991): 18.

116 *In January 2004 Goldman Sachs*: Levin, "Levin and Collins Urge Suspension of SPR Oil Deposits."

*CHAPTER SIX*

120 *In the 1970s OPEC supplied 67 percent*: Joseph G. Haubrich, Patrick Higgins, and Janet Miller, "Oil Prices: Backward to the Future?" Federal Reserve Bank of Cleveland, December 2004, p. 1.

121 *The New York Mercantile Exchange opened in 1872*: New York Mercantile Exchange Fact Sheet "Why Do They Need to Yell . . ." p. 4.

121 *In 1977 Maine farmers defaulted; The exchange's chairman called*: Samuel Glasser, "Born of Serendipity, Heating Oil Futures Spawned a Powerhouse," *Energy in the News* 3 (New York: New York Mercantile Exchange, 2003), pp. 7–8.

122 *By 1982 half the world's crude oil*: Haubrich et al., "Oil Prices."

123 *T. Boone Pickens's story*: Boone Pickens, "Oil Futures . . . A Better Brand of Risk Management," *Energy in the News*, 2003, pp. 14–15.

123 *Only 5 percent of the futures contracts; Number of paper barrels traded at the NYMEX*: Pelin Berkmen, Sam Ouliaris, and Hossein Samiei, "The Structure of the Oil Market and Causes of High Prices," IMF Research Department, (September 2005), http://www.imf.org/external/np/pp/eng/2005/092105o.htm.

127 *His research suggests that futures trading*: Robert J. Weiner, "Energy Futures Markets—Myths and Realities," *IAEE Newsletter* (4th Quarter 1999) and Robert Weiner, "Do Birds of a Feather Flock Together?: Speculator Herding in Derivative Markets," paper, George Washington University, September 2004.

129 *"You have to watch that like a hawk"*: Bhushan Bahree, et al., "Five Who Laid the Groundwork for Historic Spike in Oil Market," *Wall Street Journal*, December 20, 2005.

130 *In early 1973, before the Arab oil embargo*: David Halberstam, *The Reckoning* (New York: William Morrow, 1986), pp. 14, 15.

130 *"The new basis for doing business . . ."*: Charles T. Maxwell, "The End of August: An Interim in Oil Pricing," Weeden & Co., September 13, 2005.

133 *Saudi Arabia watched its earnings fall*: *Country Studies/Area Handbook: Saudi Arabia*, Library of Congress, accessed online at: countrystudies.us/saudi-arabia/41.html.

134 *"OPEC is much more than a non-cooperative oligopoly . . ."*: James L. Smith, "The Inscrutable OPEC: Behavioral Tests of the Cartel," *Energy Journal* 26, no. 1 (January 2005), p. 30.

135 *Congress let SPR authorization expire*: Bamberger, "Strategic Petroleum Reserve," CRS Issue Brief for Congress, August 2, 2001, p. 2.

135 *"In 1998 alone OPEC members . . ."*: Ali Rodríguez Araque speaking at the 10th Annual Middle East Petroleum and Gas Conference, Doha, Qatar, April 8, 2002.

136 *Asia was burning a million*: Anthony H. Cordesman and Khalid R. al-Rodhan, "The Changing Risks in Global Oil Supply and Demand: Crisis or Evolving Solutions?" Center for Strategic and International Studies, Working Draft, October 3, 2005, p. 36.

137  *every $10 rise in the price of a barrel of oil*: International Energy Agency, "Analysis of the Impact of High Oil Prices on the Global Economy" (2004).

137  *consumers are paying trillions of dollars*: Andrew Higgins and Gregory L. White, "Oil Producers Gain Global Clout from Big Windfall," *Wall Street Journal*, October 4, 2005.

CHAPTER SEVEN

138  *People say that Venezuela is; RR o 350*: International Crisis Group, "Latin America Briefing: Venezuela; Headed Toward a Civil War?" Quito, May 10, 2004.

138  *while in 2004 the revenues of the state of Venezuela were $26 billion*: Central Intelligence Agency, *World Factbook* (2005), accessed at: http://www.umsl.edu/services/govobcs/nofact2005/geo/ve.html.

138  *the revenue of PDVSA that year was $42 billion*: Juan Forero, "42 Billion in Sales According to SEC Filings," *New York Times*, July 24, 2004.

140  *Venezuela supplies 12 percent*: Energy Information Administration, "Country Analysis Briefs: Venezuela" (September 2005).

140  *The Venezuelan ambassador to the United States*: Interview by author with Ambassador Bernardo Alvarez Herrera, Berkeley, CA, October 2003.

140  *"Analysts . . . underestimated President Chávez's willingness . . ."*: Michelle Billig, "The Venezuelan Oil Crisis: How to Secure America's Energy," *Foreign Affairs* (September/October 2004): 3.

140  *employs fewer than fifty thousand people*: Mark P. Sullivan, "Congressional Research Report, Report for Congress, Venezuela: Political Conditions and U.S. Policy," May 18, 2005.

141  *"respect must be extended to Bolívar's statue . . ."*: Hillary Dunsterville Branch, *Venezuela: The Bradt Travel Guide*, 4th ed. (Bucks, U.K.: Bradt Travel Guides, 2003), p. 57.

142  *In 2004 PDVSA spent $1.7 billion*: Brian Ellsworth, "The Oil Company as Social Worker," *New York Times*, March 11, 2004.

142  *Between 1980 and 1999, Venezuelan's income fell*: The Federal Reserve Bank of Dallas, "The 'Curse' of Venezuela, Issue 3," *Southwest Economy* (May/June 2004).

142  *More than half the population now lives on less than $2 a day*: Franklin Foer, "The Talented Mr. Chávez," *The Atlantic Monthly* (May 2006).

143  *The country's oil arrived*: Miguel Tinker Salas, "Fueling Concern: The Role of Oil in Venezuela," *Energy* 26, no. 4 (Winter 2005), p. 1, accessible at http://hir.harvard.edu/articles/1296/.

143  *U.S. and European oil companies were pulling out; The Venezuelan state then consisted*: Terry Lynn Karl, *The Paradox of Plenty: Oil Booms and Petro-States* (Berkeley: University of California Press, 1997), pp. 77, 76.

144  *"If there had been no perfect . . ."; "oil operations are . . ."*: Miguel Tinker Salas, "Staying the Course: United States Oil Companies in Venezuela, 1945–1958," *Latin American Perspectives* 32, no. 2 (March 2005), pp. 149, 148.

144  *"America 20 years from now . . ."*: 2001 National Energy Policy, "Overview," p. 3, http://www.whitehouse.gov/energy/overview.pdf.

145  *military dictator Marcos Pérez Jiménez*: Fernando Coronil, *The Magical State* (Chicago: University of Chicago Press, 1997), p. 156.

145  *oil put $7 billion; Beer made up half of the increase*: Norman Gall, "Oil and Democracy in Venezuela, pt. 1: Sowing the Petroleum," American Universities Staff Report (January 1973): 3, 6.

*145 in 1952 Venezuelans spent $5.7 million importing eggs*: Rómulo Betancourt, *Venezuela: Oil and Politics*, trans. Everett Bauman (Boston: Houghton Mifflin, 1979), p. 305.

*146 "Here two and two make twenty-two instead of only four"*: Quoted in Rómulo Betancourt, *Venezuela's Oil*, trans. Donald Peck (London: George Allen & Unwin, 1978), p. 307.

*146 one in seven belonged*: Interview by author with Alfredo Keller, Caracas, April 2004.

*147 taxes alone don't foster accountability*: Michael L. Ross, "Does Taxation Lead to Representation?" University of California, Los Angeles Department of Political Science, January 27, 2003, p. 28.

*147 With the downturn, some governments*: Karl, *The Paradox of Plenty*, pp. 258–259, 193.

*152 Chávez bought 100,000 AK-47s*: Mike Ceaser, "Chávez's 'Citizens Militias' on the March," BBC News, July 1, 2005.

*154 Along the shores of Lake Maracaibo*: Miguel Tinker Salas, "Culture, Power and Oil: The Experience of Venezuelan Oil Camps and the Construction of Citizenship," in *Race, Class and Empire*, ed. Gilbert Gonzales (Rutledge Press, *forthcoming*).

*156 "The industry has long-term strategies*: Cesar E. Baena, *The Policy Process in a Petro-State* (Aldershot, UK, and Burlington, VT: Ashgate Publishing, 1999), pp. 159, 146, 235, 219.

*156 one of the contracts gave the state paltry royalties*: Juan Forero, "Energy-Rich Nations Are Raising Price of Foreign Admittance," *New York Times*, July 5, 2005.

*156 there were discussions about privatizing PDVSA*: Salas, "Fueling Concern," p. 3.

*157 "overcome a colonialist past"*: "Rodriguez on Turbulent Relations Between PDVSA and Venezuelan Government," *Middle Eastern Economic Survey* 47, no. 39 (September 2004).

*158 "If Mr. Bush gets the mad idea . . ."*: Patricia I. Vasquez, "Venezuelan Oil Flows Normal after Chavez Threats to US," *The Oil Daily*, March 2, 2004.

*163 The company announced that, by the end of 2004*: Peter Wilson, "PDVSA Says Social Spending Outstripped Investments (Update), Bloomberg.com, October 11, 2005.

*164 By 2006 there were even rumors:* Andy Webb-Vidal, "Venezuela Buys Russian Oil to Avoid Defaulting on Deals," *Financial Times*, April 28, 2006.

*164 By 2006 Venezuela expanded its political reach*: Juan Forero, "Chávez, Seeking Foreign Allies, Spends Billions," *New York Times*, April 4, 2006.

*164 Venezuela "has a strong card to play . . ."*: Justin Blum, "Chavez Pushes Petro-Diplomacy," *Washington Post*, November 22, 2005.

*165 The Department of Defense labeled Venezuela*: Bill Arkin, "Early Warning," *Washington Post*, November 2, 2005, citing Department of Defense briefing FY08-13PDM.

## Other Sources Used

Mommer, Bernard. *Global Oil and the Nation State* (Oxford: Oxford University Press, Oxford Institute for Energy Studies, 2002).

Padgett, Tim. "The Latin Oil Czar," *Time*, July 26, 2004.

Salas, Miguel Tinker. "Venezuelans, West Indians and Asians: The Politics of Race in Venezuelan Oil Fields 1920–1940," in *Work, Protest and Identity, Twentieth Century*

*Latin America*, ed. Vincent Peloso (Wilmington, DE: Scholarly Resources, 2003), pp. 143–164.

## CHAPTER EIGHT

*170 oil industry publications note . . .* : Paul F. Hueper, "Tapping into a New Frontier Oil Province," *Petroleum Economist* (November 1998); http://www.essochad.com/Chad/Library/News/Chad_NW_mediabis_011098.asp (accessed on 11/05/03).

*171 The first president has been described as Ubu-esque*: Sam C. Nolutshungu, *Limits of Anarchy* (Charlottesville: University Press of Virginia, 1996), p. 72.

*171 a collection of warlords fought back*: Amnesty International, "Chad: The Habré Legacy," October 16, 2001, pp. 22, 23.

*171 Déby . . . held an election in 1996*: U.S. Department of State, "Chad Country Report on Human Rights Practices for 1998," Bureau of Democracy, Human Rights, and Labor, February 26, 1999.

*173 Between 1992 and 2004, 82 percent of World Bank investments*: Jim Vallette and Steve Kretzmann, "The Energy Tug of War," Sustainable Energy & Economy Network, 2004, p. 7.

*173 In 1981, after the second oil crisis*: Vallette and Kretzmann, *Energy Tug of War*, p. 9, note 11.

*173 By 2005 Exxon's actual production*: Bhushan Bahree and Jeffrey Ball, "Oil Giants Face New Competition for Future Supply," *Wall Street Journal*, April 19, 2005.

*173 The yearly rise in non-Russian, non-OPEC oil*: Herman Franssen, "The End of Cheap Oil: Cyclical or Structural Change in the Oil Market?" *Middle Eastern Economic Survey*, February 7, 2005, p. 2.

*173 "African oil is not an end but a means . . ."; two-thirds to three-quarters of American investment*: Institute for Advanced Strategic & Political Studies symposium, "African Oil: A Priority for US National Security and African Development," January 25, 2002, p. 4.

*174 One study found that oil exporters spend*: Ian Gary and Terry Lynn Karl, "Bottom of the Barrel: Africa's Oil Boom and the Poor," Catholic Relief Services, 2003, p. 23.

*175 An Exxon executive told a reporter*: Roger Thurow and Susan Warren, "In War on Poverty, Chad's Pipeline Plays Unusual Role," *Wall Street Journal*, June 24, 2003.

*175 Exxon is permitted to import*: The World Bank and International Finance Corporation, "Project Appraisal Document Chad/Cameroon Petroleum Development and Pipeline Project," Report No. 19343 AFR, April 13, 2003, p. 25.

*175 PFC Energy, an oil and gas consulting company, estimates that Chad's take*: Ian Gary and Nikki Reisch, "Chad's Oil: Miracle or Mirage?: Following the Money in Africa's Newest Petro-State," Catholic Relief Services and Bank Information Center, February 2005, p. 39.

*175 Commercial banks reviewed Exxon's magic numbers; So Exxon asked the World Bank*: World Bank, Report No. 19343 AFR, pp. 22, 37.

*175 "Chad could not afford to lose . . ."*: Thurow and Warren, "In War on Poverty."

*175 Its report gave Chad a "Significant" rating*: World Bank, Report No. 19343 AFR, p. 37.

*175 A study by Jeffrey Sachs*: Jeffrey Sachs and Andres Warner, "Natural Resources and Economic Growth," Development Discussion Paper no. 517, Harvard Institute for International Development, 1995.

*176* **At Oxford, Paul Collier's regression studies**: Paul Collier and Anke Hoeffler, "Greed and Grievance in Civil War," Policy Research Working Paper no. 2355, Development Research Group (Washington, DC: World Bank, May 2000).

*176* **For every dollar the country got from aid in 1998**: Statistics from African Forum & Network on Debt & Development; Debt Profile for Chad, 2006, http://www.afrodad.org/.

*176* **a successful strategy for the International Finance Corporation**: Gary and Karl, "Bottom of the Barrel," p. 15.

*177* **"governance was weakening . . ."**: John M. Fitzgerald, in documents sent to the author, July 28, 2004.

*177* **Déby . . . dutifully gave $13.5 million**: Gary and Reisch, "Chad's Oil: Miracle or Mirage?" p. 57.

*177* **"It seems particularly unconscionable"**: Donald Norland, in testimony before the Subcommittee on Africa of the Committee on International Relations of the House of Representatives, April 18, 2002, p. 27, accessed at commdocs.house.gov/committees/intlrel/hfa78803.000/hfa78803_0.HTM.

*177* **And the EXIM bank**: Gary and Karl, "Bottom of the Barrel," p. 64.

*178* **"if human rights have 'significant direct economic benefits' . . ."**: Gary and Karl, "Bottom of the Barrel," p. 67, n. 41.

*179* **On July 28, 2003, a car owned by a citizen**: U.S. Department of State, "Chad Country Report on Human Rights Practices for 2003," Bureau of Democracy, Human Rights, and Labor, February 25, 2004. p. 2.

*179* **Déby has a long history**: U.S. Department of State, "Chad Country Report on Human Rights," p. 7.

*179* **But his family members appeared to be supplying arms to; Senior officers in the Chadian army**: Integrated Regional Information Networks, "Chad and the Darfur Conflict," *All Africa*, February 16, 2004.

*180* **important Salafist nicknamed "Al Para"**: Craig S. Smith, "Chad Rebel Group Says It Holds Qaeda-Linked Terrorist," *New York Times*, May 14, 2004.

*180* **Marines sent to train Chadian soldiers**: Ann Scott Tyson, "US Pushes Anti-Terrorism in Africa," *Washington Post*, July 26, 2005.

*180* **During that execution, for example**: "Un member du peloton d'exécution, blessé aux tirs, est mort par anémie," *Le Progrés*, November 10, 2003, p. 8.

*181* **Only the 1 billion barrels in the project are protected**: Gary and Reisch, Chad's Oil, pp. 29, 30, 31, and 28.

*183* **the anthropologist the company hired**: Jerry Useem, "Exxon's African Adventure," *Fortune*, April 15, 2002, pp. 102–106.

*183* **She has been quoted as saying**: Ken Silverstein, "AIDS Could Follow African Pipeline," *Los Angeles Times*, June 18, 2003, p. 1.

*184* **They called southern Chad Le Tchad-utile**: Mario Joaquim Azevedo, *Roots of Violence: A History of War in Chad* (Amsterdam: Overseas Publishers Association, 1998), pp. 75, 77.

*185* **The lure of money drained the university**: Gary and Karl, "Bottom of the Barrel," p. 98, n. 33.

*186* **"Construction freight equal to nearly 50 Eiffel Towers**: Chad/Cameroon Development Project, Exxon News Media Fact Sheet, October 2003.

*190* **By 2004 robberies and attacks in the oil region**: Emily Wax, "Oil Wealth Trickles into Chad," *Washington Post*, March 13, 2004.

*191* **In 2004 a World Bank panel recommended**: Christopher Walker, "Governing on Empty," *Barron's*, October 31, 2005.

191 *Exxon says that by mid-2003, it held 1,800 meetings with 77,000 attendees:* Chad/Cameroon Development Project.

195 *A 1997 report from Amnesty International:* Amnesty International, "AI Report 1997: Chad," Amnesty International Publications. http://www.amnesty.org/ailib/aireport/ar97/AFR20.htm

195 *Ngalaba was also the site of an ongoing medical survey:* Lori Leonard, "Possible Illnesses: Assessing the Health Impacts of the Chad Pipeline," *Bulletin of World Health Organization* (2003), pp. 427–433.

197 *Chad demanded that the company pay $50 million:* Chip Cummins, "Exxon Faces Dilemma on Chad Project," *Wall Street Journal*, February 28, 2006.

## CHAPTER NINE

201 *One American account:* Gary Stubblefield with Hans Halberstadt, *Inside the US Navy SEALs* (Osceola, FL: MBI Publishing Company, 1995), p. 165.

202 *one of the most influential naval battles:* Craig L. Symonds, *Decision at Sea: Five Naval Battles That Shaped American History* (New York: Oxford University Press, 2005), pp. 287, 288.

202 *"dealing in perceptions"; "The Gulf presented . . .":* Admiral William J. Crowe Jr., *The Line of Fire* (New York: Simon and Schuster, 1993), pp. 201, 179.

202 *"The use of sea lanes of the Gulf . . .":* Michael A. Palmer, *Guardians of the Gulf: A History of America's Expanding Role in the Persian Gulf* (New York: Free Press, 1992), p. 124.

202 *the nine-hour fight ended with . . . :* Patrick E. Tyler, "US-Iranian Naval Clash Stymies Oil Traffic in Gulf; Washington Post Foreign Service, April 19, 1988.

202 *six Iranian ships sunk:* George C. Wilson and Molly Moore, "US Sinks or Cripples 6 Iranian Ships in Gulf Battles; No American Loses Reported, but Helicopter Missing," *Washington Post*, April 19, 1988.

203 *the U.S. military presence in the Gulf; By one estimate:* Testimony of Milton R. Copulos, President, National Defense Council Foundation, before the Senate Foreign Relations Committee, March 30, 2006.

204 *the bombing was perceived as the first half of a "message":* Kenneth M. Pollack, *The Persian Puzzle* (New York: Random House, 2004), p. 232.

205 *"Everyone wants to go to Baghdad . . .":* David Remnick, "The Talk of the Town: War Without End?" *The New Yorker*, April 14, 2003.

205 *the International Court of Justice:* Pieter H. F. Bekker, "The World Court Finds that US Attacks on Iranian Oil Platforms in 1987–1988 Were Not Justifiable as Self-defense, but the United States Did not Violate the Applicable Treaty with Iran," *American Society of International Law* (November 2003).

207 *"Although the missile has been paraded with the banner . . .":* Iran News Political Desk, "Ayatollah Khamenei Watches Long-range Missile Test," *Iran News*, September 19, 2004.

207 *if Iran "threw caution to the wind,":* Dr. John Chipman, "Iran's Strategic Weapons Programmes: A Net Assessment," International Institute for Strategic Studies (IISS), Arundle House, London, September 6, 2005.

207 *"Iran's quest for nuclear weapons . . .":* Ray Takeyh, "Iran: Tehran's Nuclear Recklessness and the US Response," Council of Foreign Relations, November 15, 2005 http://www.cfr.org/publication/9263/iran.html

208 *In the 1970s the United States supported the Shah's plan:* Dafna Linzer, "Past Arguments Don't Square with Current Iran Policy," *Washington Post*, March 27, 2005.

208   *"Iran lacks technology to produce black chador"*: title of article in: *Tehran Times International Daily*, September 28, 2004.

209   *estimated that the country; "increased purchasing power . . ."*: Palmer, *Guardians of the Gulf*, pp. 37, 26.

210   *the United States went so far as to ship explosives*: Rachel Bronson, *Thicker Than Oil: America's Uneasy Partnership with Saudi Arabia* (New York: Oxford University Press, 2006), citing works by Steve Everly of the *Kansas City Star*, pp. 58, 59.

210   *"governed by irresponsible policies . . ."; "one which could govern Iran . . ."*: Dr. Donald Wilber, "CIA Clandestine Service History: Overthrowing Premier Mossadeq of Iran: November 1953–August 1954" (March 1954), accessed on the Web from National Security Archives, Electronic Briefing Book No. 28, The History of the Iran Coup, 1953; http://www.gwu.edu/~nsarchiv/NSAEBB/ NSAEBB28/#documents

210   *The Shah was an anxious man*: David Harris, *The Crisis: The President, the Prophet, and the Shah—1979 and the Coming of the Militant Islam* (New York: Little, Brown, 2004), pp. 47, 48.

210   *America should not be "sanguine"*: Committee of Energy and Natural Resources, United States Senate, 95th Congress, "Access to Oil—The United States Relationships with Saudi Arabia and Iran, No. 95-70," printed at the request of Henry M. Jackson (Washington, DC: United States Government Printing Office, December 1977), p. 106.

211   *In a country with unemployment of*: Angel M. Rabasa et al., *The Muslim World after 9/11* (Santa Monica, CA: RAND Corporation, 2004), p. 209.

212   *export crude oil and then pay to reimport gasoline*: Energy Information Administration, "Iran Analysis Brief," 2005. http://www.eia.doe.gov/emeu/cabs/Iran/ Background.html

212   *Every year five thousand people die*: BBC News, "Hundreds Treated over Tehran Smog," December 10, 2005.

214   *controlling as much as 20 percent of Iran's GDP*: Paul Klebnikov, "Millionaire Mullahs," *Forbes Magazine*, July 21, 2003; http://www.forbes.com/forbes/2003/0721/056_print.html

214   *When the Norwegian state oil company*: Bloomberg, "Rafsanjanis Are Iran's Power Broker," *Alexander's Gas and Oil Connections* (May 2004); http://www.gasandoil.com/goc/news/ntm41817.htm

214   *Most households appear to spend*: Jahangir Amuzegar, "Iran's Underground Economy," *Middle East Economic Survey*, September 8, 2003.

218   *Shortly after the war Iran's oil ministry*: Youssef M. Ibrahim, "For Ghost Oil Town in Iran, Postwar Renewal Is Starting," *New York Times*, February 17, 1989.

218   *By the mid-1980s Iran was rationing food*: "The Impact of Casualties on Society," and "War Costs," *Library of Congress Country Studies: Iran*, data as of December 1987, accessed at: http://rsb.loc.gov/frd/cs/irtoc.html

220   *UAE, with a population; 3,395 missiles*: Richard F. Grimmet, CRS Report for Congress, "Conventional Arms Transfers to Developing Nations 1997–2004," CRS-49, Table 1H: Arms Transfer Agreements with Nuclear East, by Supplier, August 29, 2005.

223   *At 8:04 they began shelling*: Symonds, *Decision at Sea*, p. 300.

223   *"a large crowd of converted martyrs . . ."*: Palmer, *Guardians of the Gulf*, p. 141.

223   *Then a U.S. helicopter*: Symonds, *Decision at Sea*, p. 301.

224   *Saudi Arabia gave Iraq*: Bronson, *Thicker Than Oil*, p. 164.

225 *"desperate to make sure that Iraq . . ."*: Patrick E. Tyler, "Officers Say US Aided Iraq in War Despite Use of Gas," *New York Times*, April 18, 2002.

225 *"Time is not on our side anymore"*: Patrick E. Tyler, "US Iranian Naval Clash Stymies Oil Traffic in Gulf," *New York Times*, April 19, 1988.

225 *Sirri is thought to be studded*: Nuclear Threat Initiative Fact Sheet, "Sirri Island," last updated March 2004, accessed at: www.nti.org/e_research/profiles/Iran/Missile/3876_3920.html.

225 *Every day, nearly 16 million barrels*: Energy Information Administration, "Persian Gulf Oil and Gas Exports Fact Sheet," September 2004.

225 *"We have told the Europeans very clearly . . ."*: Neil King, "Iran Holds Big Bargaining Chips in Dispute," *Wall Street Journal*, April 18, 2005.

227 *Inside Iran, Khomeini called the war a "blessing"*: Angel M. Rabasa, RAND, p. 226.

227 *They did not answer to the Iranian navy*: Bernard E. Trainor, "The Effect of the Attack: No Shift in Iran's Goals," *New York Times*, April 18, 1988.

227 *It's not clear whether the United States*: Symonds, *Decision at Sea*, p. 296.

228 *"given a clear demonstration . . ."*: Pollack, *The Persian Puzzle*, pp. 299–300.

228 *"shock and awe"*: Harlan K. Ullman, James P. Wade, and L. A. Edney, *Shock and Awe: Achieving Rapid Dominance* (Washington, DC: National Defense University Press, 1996), p. 46 (electronic version).

228 *In a 2002 navy war game; The mine that the USS Roberts hit*: Barry R. Posen, "Command of the Commons: The Military Foundation of US Hegemony," *International Security* 28, no. 1 (2003), p. 38, and Robert Burns, "Ex-General Says War Games were Rigged," Associated Press, August 16, 2002.

231 *In 2005 the oil minister mentioned*: Reuters, "Iran Seeking to Sign Key Oil Deal with China by January," *New York Times*, December 17, 2005.

231 *In late 2004 it agreed to sell*: Kaveh Afrasiabi, "China Rocks the Geopolitical Boat," *Asia Times*, November 6, 2004.

232 *China and Russia . . . recently agreed to sell*: Reuters, "Iran and Russia Sign $1 Billion Defense Deal: Reports," *New York Times*, December 2, 2005.

232 *"We don't care about the sanctions . . ."*: Vivian Walt, "Iran Looks East," *Fortune*, February 8, 2005.

233 *"As long as we define . . ."*: Saideh Lotfian, "A Regional Security System in the Persian Gulf, Chapter 4," in Lawrence Potter and Gary Sick, *Security in the Persian Gulf* (New York: Palgrave Macmillan, 2002), pp. 123–124.

*CHAPTER TEN*

235 *Now, on good days, Nigeria exports 2.5*: Amnesty International, "Ten Years On: Injustice and Violence Haunt the Oil Delta," October 2003, p. 3.

235 *Under the Niger Delta*: Jeffrey Taylor, "Worse Than Iraq?" *The Atlantic Monthly* (April 2006), and Michael Peel, "Crisis in the Niger Delta: How Failures of Transparency and Accountability Are Destroying the Region," Chatham House/Royal Institute of International Affairs, July 2005, p. 2.

236 *With a thousand battle-related deaths a year*: WAC Global Services, "Peace and Security in the Nigerian Delta: Conflict Expert Group Baseline Report," December 2003.

238 *"I don't engage in bunkering"*: Human Rights Watch, "Violence in Nigeria's Oil Rich Rivers State in 2004, A Human Rights Watch Briefing Paper," February 2005, p. 7.

*239*    *In Colombia, right-wing paramilitaries*: Jeremy McDermott, "Colombia cracks down on oil theft," British Broadcasting Company, February 15, 2005.

*239*    *half the area's oil is being hijacked*: Umalt Dudayev, "Special Investigation: Russians and Chechens Collude in Oil Trade," CRS No. 150, Institute for War and Peace Reporting, October 10, 2002. http://www.iwpr.net

*239*    *"oil-smuggling mafia"*: James Glanz and Robert F. Worth, "Oil Graft Fuels the Insurgency, Iraq and US Say," *New York Times*, February 5, 2005.

*239*    *two Nigerian admirals were charged*: UN Integrated Regional Information Networks, "Nigeria: Conviction of Admirals Confirms Navy Role in Oil Theft," All Africa Global Media, January 6, 2005.

*239*    *Shell's "business life expectancy"*: WAC Global Services, "Peace and Security in the Nigerian Delta," p. 79.

*240*    *A 2004 World Bank study*: Jeffrey Tayler, "Worse Than Iraq?" Atlantic Monthly, April 2006, p. 33.

*242*    *the Nigerian army showed up in boats provided by Shell*: Karl Maier, *This House Has Fallen: Nigeria in Crisis* (Middlesex, UK: Penguin Books, 2000), pp. 123–125.

*242*    *Troops raided Oloibiri*: Oronto Douglas and Ike Okonta, *Where Vultures Feast: Shell Human Rights, and Oil* (New York: Verso, 2003), p. 151.

*242*    *nearly a dozen delta communities*: Human Rights Watch, "Violence in Nigeria's Oil Rich Rivers State in 2004," p. 16, and Amnesty International, "Ten Years on: Injustice and Violence Haunt the Oil Delta," Section 4.2.

*242*    *Just two months before my arrival*: Amnesty International, "Ten Years On: Injustice and Violence Haunt the Oil Delta."

*242*    *A few days after the destruction*: Issac Olamikan, "Odiama Crisis: Judicial Commission Submit Report," *Port Harcourt Telegraph*, June 1, 2005.

*243*    *A World Bank study of delta families*: World Bank, "Nigeria: Poverty, Environment and Natural Resources Linkages," June 2003.

*243*    *Unemployment in delta villages*: Stephan Davis and Dimieari Von Kemedi, "The Current Stability and Future Prospects of Peace and Security in the Niger Delta," *Our Niger Delta*, March 1, 2006, pp. 6, 7.

*245*    *In 1996 the government asked Shell*: Omon-Julius Onaba, "Nigeria: Gas Flaring: Court Overturns Ruling Against Shell," *This Day* (Lagos), May 26, 2006.

*246*    *One official tells the Wall Street Journal*: Chip Cummins, "A Nigerian Cop Cracks Down on a Vast Black Market in Oil," *Wall Street Journal*, April 13, 2005.

*247*    *Credible rumors connect him to support of political violence*: The Small Arms Project, "The Big Disarmament Gamble; The Comeback of The Small Arms and Light Weapons," Niger Delta Project for Environment, Human Rights and Development, September 2005, p. 8.

*247*    *He was later accused of helping himself to millions*: Lydia Polgreen, "As Nigeria Tries to Fight Graft, a New Sordid Tale," *New York Times*, November 29, 2005.

*249*    *"Shell is not the government"*: Shell Inc., "Human Rights Dilemmas: A Training Supplement," produced by Visual Media Services, August 2002, p. 14.

*249*    *In 2004 Shell admitted that it had overestimated*: Jeff Gerth and Stephen Labaton, "Shell Withheld Reserves Data to Aid Nigeria," *New York Times*, March 19, 2004.

*250*    *the company was spending only $25 million*: Shell Inc., "Shell and the Environment 2004 Annual Report," 2004, pp. 35, 33.

*250*    *Shell had to break out violence into two categories*: Shell Inc., "Shell and the Environment 2005 Annual Report," 2005, pp. 4, 5.

252  *By 1996 two-thirds of the country lived in poverty*: World Bank, "Nigeria: Poverty, Environment and Natural Resources Linkages," p. 7.

254  *Asari was reportedly receiving*: International Crisis Group, "The Swamps of Insurgency: Nigeria's Delta Unrest," Africa Report no. 113, August 3, 2006, pp. 10–11. Accessed online at http://www.icg.org.

255  *the regions rigged elections*: Peel, "Crisis in the Niger Delta," p. 3.

257  *the effect at New York's NYMEX was huge*: Energy Intelligence Group, "Nigeria: Gangsta Power," *Energy Compass*, October 1, 2004, and Bronwyn Manby, "Oil Jihad in the Niger Delta," openDemocracy.org, October 7, 2004.

260  *Later he denied saying this*: Charles Ozeman, "Presidency Denies Atiku," *The Vanguard*, March 2, 2006.

261  *In 2005 China invested $3 billion*: Dino Mahtani, "A Glimpse of the World: Nigeria Shifts to China Arms," *Financial Times*, March 1, 2006, and Peter S. Goodman, "CNOOC Announces $2.3 B Nigeria Investment," *Washington Post*, January 9, 2006.

262  *Nigeria was able to cut a deal*: "Clean Slate," *The Economist*, October 20, 2005; http://www.economist.com/displayStory.cfm?story_id=5065583.

263  *Obasanjo appeared at the World Economic Forum*: Josephine Lohor, "We'll Make Nigeria World Tourist Destination," *This Day* (Lagos), January 20, 2006.

263  *By 2006, as violence worsened*: Josephine Lohor, "FG to Commit N20 Trillion," *This Day* (Lagos), April 19, 2006.

### Other Sources Used

Althaus, Dudley. "Politics of Oil Inflame Age-Old Delta Hatreds," *Houston Chronicle*, December 17, 2004.

Apter, Andrew. *The Pan-African Nation: Oil and the Spectacle of Culture in Nigeria* (Chicago: University of Chicago Press, 2005).

Donnelly, John. "Burdens of Oil Weigh on Nigerians: Ecological Harm, Corruption Hit Hard," *Boston Globe*, October 3, 2005.

Okonta, Ike. "Nigeria: Chronicle of a Dying State," *Current History* (May 2005): 206.

Reno, William. *Warlord Politics and African States* (Boulder, CO: Lynne Rienner Publishers, 1998).

Ross, Michael L. "Nigeria's Oil Sector and the Poor," UCLA Department of Political Science, May 23, 2003.

*CHAPTER ELEVEN*

265  *Statistics on car ownership; has the same number of cars per capita*: author interview with Dr. Lee Schipper, of EMBARQ with the World Resources Institute, telephone, February 2005.

265  *In 2005 the electrical situation was on the mend*: Patrick Barta, "China Continues to Fill Up on Oil, but Pace Slackens," *Wall Street Journal*, November 9, 2005, and Don Lee, "China Making Big Oil Moves," *Los Angeles Times*, January 23, 2006.

266  *"We have been producing and exporting oil for more than 100 years"*: Quoted in Juan Forero, "China's Oil Diplomacy in Latin America," *New York Times*, March 1, 2005.

269  *U.S. hegemony is particularly useful for China*: Jonathan Story, "The Global Implications of China's Thirst for Energy," *Middle Eastern Economic Survey*, February 16, 2004.

269 *In 2004 the country shook automakers worldwide*: Feng An, Amanda Sauer, and Fred Willington, "Taking the High (Fuel Economy) Road," *World Resources Institute* (November 2004), p. 1.

270 *China's greenhouse gas emissions*: "China to Cut Energy Use by Four Percent in 2006," *People's Daily*, March 5, 2006.

270 *But by late 2005*: "Nation Lifts Target for Renewable Energy Use," *China Daily*, November 7, 2005.

271 *Premier Wen Jiabao announced*: "China to Cut Energy Use."

272 *For one thing, pollution may be choking off*: Warwick J. McKibbin, "Environmental Consequences of Rising Energy Use in China," Brookings Institution and the Lowy Institute for International Policy, Sydney, revised December 10, 2005, p. 7.

273 *the process of converting natural gas to hydrogen*: Pamela L. Spath and Margaret K. Mann, "Natural Gas Steam Methane Reforming" (Golden, Colorado: National Renewable Energy Laboratory, 2001), NREL/TP-570-27637, executive summary, p. 2.

277 *In 2000 a few companies began selling*: Phone interview by author with Jonathan Weinart of the University of California, Davis, April 2006.

## Other Sources Used

Bradsher, Keith. "China's Factories Aim to Fill the World's Garages," *New York Times*, November 2, 2003.

"Dream Machines," *The Economist*, June 2, 2005.

Jianhai, Bi, and David Zweig. "China's Global Hunt for Energy," *Foreign Affairs* 84, no. 5 (September 2005, pp. 25–38).

# INDEX

*ABOUT THE AUTHOR*

Lisa Margonelli is currently an Irvine Fellow at the New America Foundation. She has written for the *San Francisco Chronicle, Wired, Business 2.0, Discover,* and *Jane,* and was the recipient of a Sundance Institute Fellowship and an excellence in journalism award from the Northern California Society of Professional Journalists. She is based in Oakland, California.

*A NOTE ON THE TYPE*

The text of this book is set in Goudy, a typeface based on Goudy Old Style designed by the American type designer Frederic W. Goudy in 1915 for American Type Founders. Versatile enough as both text and display, it is one of the most popular typefaces ever produced, frequently used for packaging and advertising. Its open letterforms make it a highly legible and friendly text face.